W9-BSA-217

MOLECULAR
BIOLOGY
INTELLIGENCE
UNIT

Effects of Extracellular Adenosine and ATP on Cardiomyocytes

Amir Pelleg, Ph.D.
Allegheny University of the Health Sciences
Departments of Medicine, Physiology, Biophysics
and Pharmacology
Philadelphia, Pennsylvania, U.S.A.

Luiz Belardinelli, M.D.
University of Florida
Departments of Medicine and Pharmacology
Gainesville, Forida, U.S.A.

R.G. LANDES
COMPANY

AUSTIN, TEXAS
U.S.A.

MOLECULAR BIOLOGY INTELLIGENCE UNIT

Effects of Extracellular Adenosine and ATP on Cardiomyocytes

R.G. LANDES COMPANY
Austin, Texas, U.S.A.

U.S. and Canada Copyright © 1998 R.G. Landes Company

Please address all inquiries to the Publishers:
R.G. Landes Company, 810 South Church Street, Georgetown, Texas, U.S.A. 78626
Phone: 512/ 863 7762; FAX: 512/ 863 0081

U.S. and Canada ISBN: 1-57059-524-0

Library of Congress Cataloging-in-Publication Data

Effects of extracellular adenosine and ATP on cardiomyocytes / [edited by] Amir Pelleg, Luiz Belardinelli.
    p. cm. -- (Molecular biology intelligence unit)
    Includes bibliographical references and index.
    ISBN 1-57059-524-0 (alk. paper)
    1. Heart cells. 2. Adenosine--Physiological effect. 3. Adenosine triphosphate--Physiological effect. 4. Heart--Physiology.
I. Belardinelli, Luiz. II. Series.
    [DNLM: 1. Myocardium--metabolism. 2. Heart--drug effects.
3. Adenosine--pharmacology. 4. Adenosine--metabolism. 5. Adenosine Triphosphate--pharmacology. 6. Adenosine Triphosphate--metabolism.
WG 280 E27 1998]
QP114.C44E44 1998
612.1'7--dc21
DNLM/DLC   97-46653
for Library of Congress   CIP

PUBLISHER'S NOTE

R.G. Landes Company produces books in six Intelligence Unit series: *Medical, Molecular Biology, Neuroscience, Tissue Engineering, Biotechnology* and *Environmental*. The authors of our books work directly with the physical systems they describe and contribute with avid interest to progressing knowledge about these systems. Typically, topics are fascinating pieces of important biological puzzles; often few or no similar books exist on these topics.

Our goal is to publish books in important and rapidly changing areas of bioscience for sophisticated researchers and clinicians. To achieve this goal, we have accelerated our publishing program to conform to the fast pace at which information grows in bioscience. We aim to publish our books within six months or less of receipt of the manuscript. We thank our readers for their continuing interest and welcome any comments or suggestions they may have for future books.

Judith Kemper
Production Manager
R.G. Landes Company

CONTENTS

EDITORS

Amir Pelleg, Ph.D.
Allegheny University of the Health Sciences
Departments of Medicine Physiology, Biophysics
and Pharmacology
Philadelphia, Pennsylvania, U.S.A.
Introduction, Chapter 6

Luiz Belardinelli M.D.
University of Florida
Department of Medcine
Gainesville, Florida, U.S.A.
Introduction, Chapter 3

CONTRIBUTORS

Guoping Cai, M.D.
Department of Pharmacology
MCP-Hahnemann
  School of Medicine
Allegheny University
  of the Health Sciences
Philadelphia, Pennsylvania, U.S.A.
Chapter 5

Ulrich K.M. Decking, M.D.
Institut für Herz
  und Kreislaufphysiologie
Heinrich-Heine Universität
  Düsseldorf
Düsseldorf, Germany
Chapter 2

Donn M. Dennis, M.D.
Department of Anesthesiology
University of Florida
Gainesville, Florida, U.S.A.
Chapter 3

Andreas Deussen, M.D.
Institut für Physiologie
  und Pathophysiologie
Medizinische Fakultät
  Carl Gustav Carus
Dresden, Germany
Chapter 2

Eitan Friedman, Ph.D.
Department of Pharmacology
MCP-Hahnemann
  School of Medicine
Allegheny University
  of the Health Sciences
Philadelphia, Pennsylvania, U.S.A.
Chapter 5

Robert D. Lasley, Ph.D.
University of Kentucky
  College of Medicine
Department of Surgery
Lexington, Kentucky, U.S.A.
Chapter 4

Mark E. Olah, Ph.D.
Division of Cardiology
Department of Medicine
Duke University Medical Center
Durham, North Carolina, U.S.A.
Chapter 1

M.J. Pekka Raatikainen, M.D., Ph.D.
Department of Medical
  Biochemistry and Internal
  Medicine
University of Oulu
Oulu, Finland
Chapter 3

Jürgen Schrader, M.D.
Institut für Herz
  und Kreislaufphysiologie
Heinrich-Heine Universität
  Düsseldorf
Düsseldorf, Germany
Chapter 2

Gary L. Stiles, M.D.
Division of Cardiology
Department of Medicine
Duke University Medical Center
Durham, North Carolina, U.S.A.
Chapter 1

Guy Vassort, Ph.D.
Inserm &-390
Cardiovascular Pathophysiology
Montpellier, France
Chapter 6

Hoau-Yan Wang, Ph.D.
Department of Pharmacology
MCP-Hahnemann
  School of Medicine
Allegheny University
  of the Health Sciences
Philadelphia, Pennsylvania, U.S.A.
Chapter 5

PREFACE

This book deals with the regulatory role of extracellular adenosine and adenosine 5'-triphosphate (ATP) in the heart. In 1929, Drury and Szent-Gyorgyi published their classic paper in the Journal of Physiology (London) in which they demonstrated for the first time that extracellular adenosine and ATP exert specific effects in the mammalian heart independent of their role in intracellular metabolism and energetics. Subsequent studies since then have established the role of adenosine and ATP as local physiologic regulators of various cardiac functions, the receptor subtypes and the signal transduction pathways which mediate the actions of these two purine compounds.

This book covers several well studied aspects of the actions of extracellular adenosine and ATP on cardiac myocytes; from receptors to cardioprotection. Efforts were made to present comprehensive and updated text in a didactic format.

The Editors thank the Contributors and the Publisher for their efforts to ensure high quality text while complying with a tight production schedule.

CHAPTER 1

Adenosine Receptor-Mediated Signal Transduction

Mark E. Olah and Gary L. Stiles

INTRODUCTION

The array of physiologic effects produced by adenosine in the cardiovascular system and throughout the body result from the activation of cell surface proteins known as adenosine receptors (ARs). To date, four AR subtypes, A_1AR, $A_{2A}AR$, $A_{2B}AR$ and A_3AR have been cloned from many species including humans.[1,2] ARs belong to the superfamily of G protein coupled receptors and display the structural and functional features characteristic of these signaling molecules. AR subtypes differ to varying degrees in regard to tissue distribution, structure, ligand binding properties and the functional effects resulting from their activation.[1,2] This last property, the signal transduction pathways activated by AR subtypes, is the subject of this chapter. Focus is placed on topics in adenosine receptor signal transduction which have developed over the last five years with particular emphasis on the basic mechanisms of intracellular signaling. Where appropriate, associations are made between the described signaling events and important physiologic responses. The chapter concludes with a discussion of the

Effects of Extracellular Adenosine and ATP on Cardiomyocytes,
edited by Amir Pelleg and Luiz Belardinelli. © 1998 R.G. Landes Company.

information available regarding the structural features of ARs and G proteins that are known to be involved in intracellular signaling. For broader coverage of the physiologic actions of adenosine, several additional reviews are available.[1-6]

ADENOSINE RECEPTORS AND
SIGNAL TRANSDUCTION PATHWAYS

Provided below is a review, categorized by AR subtype, of the signal transduction pathways through which adenosine may produce its effects. For the most part, two means are available to identify a specific receptor subtype in mediating a particular response. First, in cells or tissues natively expressing ARs, the response may be classified pharmacologically, i.e., defined by a potency order for both stimulation by agonists and inhibition by antagonists. Secondly, the response may be demonstrated in a cell line in which the recombinant receptor is heterologously expressed. Caution must be exercised with either of these approaches. In some reports, responses have been defined as occurring due to activation of a particular AR subtype based on the affinity of a single agonist or sensitivity to a single antagonist. In general, a much more rigorous pharmacological characterization is necessary. Furthermore, species differences in the ligand binding properties of a particular AR subtype may also be relevant. Additionally, as will be described below, certain AR-mediated responses may be cell specific as they depend on the complement of signaling molecules in the cell and findings in one cell type may not be extrapolated to another. In transfected cells, factors such as receptor overexpression must be considered when assigning physiologic relevance to an observed response.

A_1AR

Signal transduction mediated by the A_1AR is for the most part typical of that demonstrated by receptors coupled to members of the $G_{i/o}$ family. The most extensively examined signaling pathway that ensues A_1AR activation is that of inhibition of adenylyl cyclase. It was this functional response which was initially employed to classify adenosine receptors, as the A_1AR was recognized to be distinct from the A_2AR, which stimulated adenylyl cyclase.[7,8] Typically demonstrated by an attenuation of enzyme activity following its stimulation by forskolin or a G_s-coupled receptor, inhibition of adenylyl cyclase by the A_1AR has been observed in freshly isolated tissues or cells,[9,10] cell lines natively expressing the A_1AR[10,11] and more recently, cells heterologously expressing various species homologues of the A_1AR.[13,14] A_1AR-mediated inhibition of adenylyl cyclase is sensitive to pertussis toxin thus implicating a G_i-dependent coupling mechanism.[10-12] However, the precise identities of the key signaling molecules have not been extensively documented and is likely in large part dependent on the cell type under examination. For example, type V and type VI adenylyl cyclases are the predominant isoforms present in the heart and both are equally sensitive to inhibition by all three $G_{\alpha i}$ isoforms but are not regulated by $G_{\beta\gamma}$ subunits.[15,16] Conversely, in the brain, type I adenylyl cyclase is present and this isoform is susceptible to inhibition by $G_{\beta\gamma}$.[15,16] As will be discussed, the A_1AR appears to couple rather efficiently to all $G_{\alpha i}$ isoforms. Thus, the precise signal transduction mechanism may be dependent on the relative population of $G_{\alpha i}$ isoforms and adenylyl cyclase isoforms present in a given cell. However, presently undefined levels of complexity may also exist as suggested by analysis of A_1AR-mediated adenylyl cyclase inhibition in rabbit intestinal smooth muscle cells.[17] These cells apparently express multiple $G_{\alpha i}$ isoforms, yet adenylyl cyclase inhibition elicited by an A_1AR agonist was antagonized specifically by antisera raised against $G_{\alpha i-3}$ but not by antisera specific for $G_{\alpha i-2}$ and $G_{\alpha i-1+2}$.[17] This isoform specific response may reflect regulation of signaling by factors such as coactivation of additional second messenger systems or confinement of specific receptor-G protein-effector interactions to distinct microdomains of the cell membrane.

Inhibition of adenylyl cyclase by the A_1AR is apparently responsible for the well-described "anti-adrenergic" effect of adenosine in the heart.[18,19] A_1AR-mediated reduction in the stimulation of adenylyl cyclase elicited by compounds such as isoproterenol attenuates the increase in calcium current elicited by the

β adrenergic receptor agonist.[3] Similarly, in cells such as adipocytes or smooth muscle cells, A_1AR-mediated inhibition of adenylyl cyclase may be of most physiologic importance following activation of the enzyme by stimulants such as catecholamines.

It is becoming increasingly apparent that signal transduction initiated by receptor-stimulated dissociation of the G protein heterotrimer is not solely dependent on the activity of released G protein α subunits. Rather, the liberated G protein $\beta\gamma$ dimeric complex may also act as an intracellular signaling molecule with downstream target effectors including adenylyl cyclase, ion channels, phospholipase C β and phosphoinositide-3 kinase.[20] The involvement of $G_{\beta\gamma}$ in A_1AR-mediated signal transduction has not been extensively examined directly. However, based on the study of receptors similarly coupled to G_i, paradigms for A_1AR signal transduction may be proposed. For example, the realization that $G_{\beta\gamma}$ may act as an intracellular effector evolved from the observation that $G_{\beta\gamma}$ could directly activate a cardiac potassium channel that opened in response to muscarinic receptor stimulation.[21] The A_1AR-mediated activation of a cardiac potassium channel with electrophysiologic properties indistinguishable from that regulated by the muscarinic receptor[22] is likely to occur via a similar $G_{\beta\gamma}$-dependent mechanism.

A second signaling cascade initiated by the A_1AR that likely involves mediation by $G_{\beta\gamma}$ is that of phospholipase C (PLC) activation with the resulting generation of inositol phosphates and increase in intracellular calcium levels. A_1AR agonists have been shown to stimulate PLC activity in cells natively expressing the receptor[23-25] as well as in Chinese hamster ovary (CHO) cells in which the recombinant A_1AR has been stably expressed.[26,27] This response to A_1AR activation has been shown to be sensitive to pertussis toxin treatment thus implicating mediation by a member of the $G_{\alpha i}$ family and not $G_{\alpha q/11}$.[23,25-27] Consistent with coupling to $G_{\alpha i}$ rather than $G_{\alpha q/11}$ is the observation that PLC stimulation by the A_1AR is typically less robust than that elicited by receptors known to couple to $G_{\alpha q/11}$. Additionally, Fredholm and coworkers[24] have demonstrated in DDT_1 MF-2 smooth muscle cells that simultaneous activation of the A_1AR and $G_{q/11}$-coupled receptors by maximal concentrations of agonist results in a more than additive stimulation of PLC, again suggesting distinct pathways between receptor and PLC. The mechanism for this synergistic response that has also been described for other cell types[28,29] is not known.

Pertussis toxin, sensitive activation of PLC is believed to occur strictly through the action of free $G\beta\gamma$ subunits.[20] This is supported by the finding that inositol phosphate accumulation stimulated by the G_i-coupled M_2- muscarinic receptor, but not that elicited by the $G_{q/11}$-coupled α_1 adrenergic receptor, is abolished by cotransfection of cells with a minigene expressing a peptide known as βARKct.[30] βARKct, representing the C-terminal 195 amino acids of the β adrenergic receptor kinase (βARK), acts as an intracellular sequestrant of free $G\beta\gamma$ subunits.[30] To date no such study of A_1AR-mediated PLC stimulation has been conducted. However, examination of PLC β activation by A_1AR agonists in rabbit intestinal circular muscle cells by Murthy and Makhlouf[17] has provided evidence for $G_{\beta\gamma}$-mediated signaling. In this system, PLC activation in response to A_1AR agonists was nearly abolished by pretreatment of permeabilized cells or cell membranes with antisera raised against the Gβ subunit.[17] Furthermore, phosphoinositide turnover following A_1AR activation appeared to occur due to stimulation of specifically PLC β_3 and not other isoforms of the enzyme.[17] This is consistent with the demonstration in reconstituted systems that purified $G_{\beta\gamma}$ preferentially activates PLC β_3.[31] Conversely, in rabbit intestinal circular muscle cells, inhibition of adenylyl cyclase by the A_1AR was not sensitive to G_β antisera but was greatly attenuated by $G_{i\alpha-3}$ antibody.[17] A puzzling observation in this study was the finding that A_1AR-mediated PLC stimulation was also blocked by $G_{i\alpha-3}$ antisera.[17] As mentioned above, $G_{i\alpha}$ subunits are not believed to activate PLC isoforms thus suggesting that A_1AR-G protein interaction may be disrupted by the $G_{i\alpha-3}$ antisera.

Activation of PLC by the A_1AR may be of physiologic importance in the phenomenon of ischemic preconditioning. It is well established that brief periods of ischemia can protect the cardiac myocyte from injury during subsequent prolonged ischemic insult.[32] The protective effects of preconditioning can be blocked by AR antagonists and mimicked by AR agonists indicating a critical role for ARs in this process.[32] Whether ischemic preconditioning involves activation of the A_1AR or A_3AR or possibly both subtypes is currently under extensive examination and may well be species dependent. One postulated mechanism of A_1AR-mediated signal transduction producing cellular protection during ischemia is a G_i-mediated stimulation of PLC resulting in activation of protein kinase C (PKC) with subsequent phosphorylation and opening of K_{ATP} channels, though variations of this model have been proposed.[32-34] It may be speculated that the ultimate activation of K_{ATP} channels results in shortening of action potential duration which reduces calcium overload and cellular energy consumption.[33] However, direct evidence supporting several of the proposed steps in this cascade is lacking and an understanding of the involved mechanisms is far from complete. Few published reports have specifically addressed the mechanistic role of the A_1AR in the initial events in this process. In rabbits, reduction of infarct size elicited by treatment with the AR agonist, R-PIA, prior to hypoxia/reperfusion was completely blocked by coadministration of the PKC inhibitors, staurosporine and polymyxin B.[35] With a study of this nature employing the intact animal, conclusions must be tempered based on possible nonspecific actions of the inhibitory compounds. Additionally, it has been demonstrated in isolated rat ventricular myocytes that R-PIA induces a translocation of the PKC-δ isoform from a cytosolic to membrane fraction of the cell which is indicative of enzyme activation.[36] Much work is needed to expand these findings and further delineate the precise steps in the A_1AR-initiated intracellular signaling process that are responsible for cardiac preconditioning. This undertaking may be complicated by factors such as the possible involvement of multiple AR subtypes in preconditioning and species dependent variations at each level of the signal transduction cascade.[31,33]

The discussion provided above summarizes what is known about A_1AR signal transduction via PLC activation. However, it should be noted that A_1AR-mediated activation of PLC has been described for a limited number of cell types. A_1AR agonists have been reported to have no effect on PLC activation when administered alone[28,37] and to inhibit PLC activity stimulated by agonists of $G_{q/11}$-coupled receptors.[12,38] In certain instances, the interpretation of these findings may be complicated by the study of tissues possessing multiple AR subtypes and the use of nonselective AR ligands. However, inhibition of PLC activity by the A_1AR in rat pituitary GH_3 cells was well described.[12]

A_3AR

Like the A_1AR, the A_3AR has thus far been documented as coupling exclusively to the G_i class of G proteins. However, in contrast to the A_1AR, the best characterized signaling pathway activated by the A_3AR is that of PLC stimulation. Not only has PLC stimulation been demonstrated in several cell types and tissues that natively express the A_3AR, but it has been associated with functional responses.

The existence of the A_3AR was not appreciated prior to its cloning in 1992[39] and thus study of signal transduction mediated by this receptor has lagged behind that of other AR subtypes. Study of the A_3AR, particularly in native cells expressing multiple AR subtypes, has also been hindered by the lack of receptor subtype selective agonist and antagonist ligands. These pharmacological considerations are also complicated by the rather markedly distinct ligand binding properties of the species homologues of the A_3AR.[39-41] However, the recent introduction of potent and at least partially selective A_3AR agonists such as IB-MECA and Cl-IB-MECA has substantially aided the study of the A_3AR.[42,43] Likewise, development of A_3AR selective antagonists is also proceeding.[44]

The A_3AR was first identified as a G_i-coupled receptor following expression of the cloned rat A_3AR in CHO cells and the demonstration of a pertussis toxin sensitive inhibition of adenylyl cyclase.[39] Inhibition of adenylyl cyclase activity by the recombinant sheep[40] and human[41] A_3ARs was subsequently shown. Interestingly, in regard to tissues or cells natively expressing the receptor, an inhibition of adenylyl cyclase mediated by the A_3AR has thus far only been reported for isolated chick ventricular myocytes.[45] Conversely, Abbracchio and coworkers[43] observed negligible if any inhibition of forskolin-stimulated adenylyl cyclase activity by A_3AR agonists in rat brain despite substantial stimulation of PLC by these compounds in this preparation.

That the A_3AR could activate PLC was initially recognized when the previously characterized atypical pharmacological profile of the AR-mediated inositol phosphate formation in rat RBL-2H3 cells was associated with the ligand binding profile described for the cloned rat A_3AR, i.e., a distinct agonist potency order and lack of sensitivity to xanthine antagonists.[46,47] Subsequently, increases in intracellular calcium levels or accumulation of inositol phosphates in rat striatal slices,[43] a mouse T lymphocyte cell line,[48] human promyelocytic HL-60 cells[49] and human eosinophils[50] have been associated with activation of the A_3AR. A_3AR signaling via stimulation of PLC was abolished by pertussis toxin in RBL-2H3 cells,[46] but a pertussis toxin insensitive response has also been reported.[48]

Signal transduction through a PLC-dependent pathway following A_3AR activation may have significant physiologic implications throughout the body. Based on the A_3AR-mediated enhancement of antigen-induced secretion in the rat mast cell model, RBL-2H3, the A_3AR has been proposed to have a role in mediating allergic responses and asthma.[51] Furthermore, Fozard and coworkers[52] have associated the A_3AR-induced release of mediators from mast cells with the hypotensive response observed upon AR agonist administration in the intact animal. Similarly, Shepherd et al[53] demonstrated that an A_3AR-mediated mast cell degranulation may regulate vasomotor tone in the hamster cheek pouch model. Interestingly, mast cell activation in this system was associated with vasoconstriction with histamine and thromboxane being possible mediators of this response.[53] The study of Meade and coworkers[54] employing the BDE rat strain implicated involvement of an A_3AR-mediated mast cell activation in the bronchospasm induced by AR agonists in these animals. The release of mediators by primed mast cells following A_3AR activation is believed to occur subsequent to the increase in intracellular calcium levels, however, the complete signal transduction pathway has yet to be delineated.

An additional physiologic role for an A_3AR-PLC signaling pathway may exist in murine CTLL-2 cells.[48] In these lymphocytic cells, activation of the A_3AR is apparently responsible for the enhancement of interleukin-2 dependent cell proliferation observed in response to AR agonists.[48] The proliferative response to AR agonists was blocked by the PLC inhibitor, D609, as well as the PKC inhibitor, staurosporine.[48] Finally, as noted previously, it appears in certain model systems that activation of the cardiac A_3AR may protect the myocyte from ischemic damage.[55,56] In conscious rabbits, Auchampach and coworkers[56] demonstrated that IB-MECA, at a concentration apparently selective for the A_3AR, protected the heart from injury induced by a myocardial stunning protocol of brief, repetitive periods of ischemia/reperfusion. This protection was completely abolished by administration of the specific PKC inhibitor, chelethyrine, prior to IB-MECA treatment, thus strongly implicating a PKC-dependent mechanism.[56]

$A_{2A}AR$

Activation of the $A_{2A}AR$ has been classically associated with a G_s-coupled stimulation of adenylyl cyclase and the resulting increase in intracellular cAMP levels.[7,8] Certain physiologic responses elicited by $A_{2A}AR$ activation may result from the stimulation of adenylyl cyclase including inhibition of platelet activity[58] and modulation of certain neutrophil functions[59,60] though direct evidence is lacking. It has also been proposed that activation of the $A_{2A}AR$ in neutrophils may produce responses such as inhibition of stimulated O_2^- generation independent of cAMP accumulation.[61] Similarly, the involvement of cAMP generation in $A_{2A}AR$-mediated vasodilation is unclear. For example, Cushing et al[62] demonstrated that both NECA and the relatively selective $A_{2A}AR$ agonist, CGS21680, relaxed endothelium-denuded bovine coronary artery preparations that had been precontracted with prostaglandin $F_{2\alpha}$. However, neither agonist significantly stimulated cAMP production in these preparations, suggesting a cAMP-independent mechanism.[62] Conversely, data from smooth muscle cells isolated from rabbit mesenteric artery support at least a partial role for cAMP in the signal transduction cascade linking $A_{2A}AR$ activation to vasodilation.[63] In these cells, CGS21680 activated an ATP-sensitive (K_{ATP}) current that was sensitive to blockade by glibenclamide.[63] This response may correlate with the adenosine-mediated glibenclamide-sensitive vasodilation of certain vascular beds.[64] Importantly, the $A_{2A}AR$-mediated K_{ATP} channel activation was attenuated by two protein kinase A inhibitors, H-89 and (R_p)-cAMP-S.[63] These electrophysiological data from isolated arterial cells suggest a paradigm of $A_{2A}AR$-mediated adenylyl cyclase stimulation resulting in cAMP-dependent PKA activation. Phosphorylation of K_{ATP} channels would result in channel opening with an ensuing membrane hyperpolarization and closing of voltage-dependent calcium channels.[63]

In regard to the mechanism of $A_{2A}AR$-mediated vasodilation, a consideration of the role of the endothelium in the intact vascular preparation adds an additional degree of complexity. In porcine coronary artery, a substantial component of the NECA- and CGS21680-induced vasorelaxation is dependent on the presence of an intact endothelium.[65] The sensitivity of the $A_{2A}AR$ agonist-induced vasorelaxation in endothelium-intact preparations to agents which prevent nitric oxide accumulation indicates the involvement of endothelial released nitric oxide acting on vascular smooth muscle.[65] The ability of NECA to stimulate an increase in intracellular cGMP content in endothelium-intact but not in endothelium-denuded preparations is consistent with this proposal.[65] An obvious focus of future study may be the identification of the signal transduction mechanism coupling $A_{2A}AR$ activation to the apparent nitric oxide release in endothelial cells.

Over the last 2-3 years, several studies employing various preparations have more directly demonstrated that certain $A_{2A}AR$-mediated responses do not occur exclusively via activation of the G_s-adenylyl cyclase-PKA system. These studies vary to the degree to which these rather novel signal transduction pathways are delineated, and focus has been placed at various levels of the signaling cascades. The following discussion primarily categorizes these studies by identity of the involved G protein and signal transduction molecules.

In isolated fetal chick ventricular myocytes, $A_{2A}AR$-mediated responses apparently occur via a G_s-dependent, yet cAMP-independent mechanism. In these cells, application of CGS21680 produced a verapamil-sensitive influx of extracellular calcium and a positive inotropic response.[66] The PKA inhibitor, (R_p)-cAMP-S, minimally inhibited the ability of CGS21680 to elicit either response, suggesting a negligible role for cAMP in the involved signal transduction pathway.[66] Similarities between the calcium influx elicited by CGS21680 and cholera toxin, e.g., only partial sensitivity to PKA inhibitors, suggest that the $A_{2A}AR$ response is mediated through G_s though direct evidence for this coupling is lacking.[66] It has been established that activation of the β adrenergic receptor can produce opening of cardiac calcium channels through a G_s-dependent, cAMP-independent pathway.[67] However, in the chick myocyte, the β adrenergic receptor- and $A_{2A}AR$-mediated increases in calcium influx and contractile amplitude appeared to occur via distinct mechanisms as maximally effective concentrations of isoproterenol and CGS21680 produced additive effects.[66] Additionally, the response to isoproterenol was completely blocked by (R_p)-cAMP-S.[66] Influx of calcium in response to $A_{2A}AR$ activation has recently been associated with an antagonism of the A_1AR- and A_3AR-mediated cellular protection that occurs during cardiac preconditioning.[45] Thus, in this system, a G_s-dependent signaling cascade that does not involve adenylyl cyclase-PKA apparently exists and has important physiologic implications.

Several recent studies indicate that activation of the $A_{2A}AR$ may initiate a signal transduction cascade that does not involve G_s. As might be expected, coupling of the $A_{2A}AR$ to a G protein other than G_s appears to result in stimulation of signaling molecules other than adenylyl cyclase. Rat pheochromoctyoma PC12 cells that natively express the $A_{2A}AR$ represent an interesting system to study such coupling. It should first be noted that it has been well established that activation of the $A_{2A}AR$ in PC12 cells produces a robust stimulation of adenylyl cyclase.[68,69] However, a dual coupling of the $A_{2A}AR$ to both G_s and G_i may exist in these cells. Koizumi et al[70] demonstrated that CGS22942, a relatively selective $A_{2A}AR$ agonist, potentiated both the increase in intracellular calcium and dopamine release elicited by ATP.[70] The potentiation was not mimicked by application of dibutyryl cAMP, indicating no involvement of the cAMP-PKA pathway.[70] Importantly, responses were abolished in cells pretreated with pertussis toxin, suggesting a role for the $G_{i/o}$ class of G proteins in coupling the $A_{2A}AR$ to the increase in intracellular calcium.[70]

In an extensive examination of $A_{2A}AR$ signaling and regulation, Lai and coworkers[71] also obtained findings that strongly suggest the $A_{2A}AR$ may couple to pertussis toxin sensitive G protein(s) in PC12 cells. In these cells, desensitization of the $A_{2A}AR$-mediated adenylyl cyclase stimulation that develops in response to continued $A_{2A}AR$ agonist exposure appears to occur at least partially due to a phosphorylation and subsequent inhibition of the type VI isoform of adenylyl cyclase.[71] Additional findings suggested that the phosphorylation of type VI adenylyl cyclase was mediated by a novel (calcium-insensitive) isoform of PKC.[71] Importantly, treatment of PC12 cell membranes with CGS21680 resulted in the phosphorylation of a PKC substrate peptide.[71] Finally, in membranes obtained from PC12 cells pretreated with pertussis toxin, the desensitization of $A_{2A}AR$-stimulated adenylyl cyclase activity in response to a 30 minute CGS21680 exposure was nearly abolished.[71] Thus, in PC12 cells, signal transduction via dual pathways may occur upon activation of the $A_{2A}AR$. It may be proposed that $A_{2A}AR$ coupling to both G_s and G_i results in adenylyl cyclase and PKC stimulation respectively, with a feedback regulatory system in existence. PC12 cells should serve as an interesting system to study not only the mechanism of activation of PKC via the $A_{2A}AR$ but also the interactions and cross-talk between two distinct signal transduction pathways. As described below, such signaling may be of physiologic relevance in neurons.

Additional studies, while not specifically identifying the involved G protein, have provided evidence that the $A_{2A}AR$ may ultimately produce certain responses via activation of PKC. Gubitz et al[72] have provided evidence that in preparations of rat striatal nerve terminals, activation of the $A_{2A}AR$ may potentiate potassium-evoked acetylcholine release via stimulation of PKC. Similar to the situation in PC12 cells, in synaptic membranes the $A_{2A}AR$ appears to display dual coupling to both PKA- and PKC-dependent pathways. Stimulation of either protein kinase appears to augment acetylcholine release via stimulation of calcium influx.[72] However, based on selective inhibition of this response by distinct calcium channel blockers, it appears the PKA-dependent pathway and PKC-dependent pathway are linked to activation of P-type and N-type calcium channels, respectively.[72] The G protein coupling the $A_{2A}AR$ to stimulation of protein kinase C was not identified, however, it does not appear to be G_s based on distinctions between the $A_{2A}AR$-mediated response and that produced by cholera toxin.[72] Rather, the findings suggest that the G_s-coupled, PKA pathway may not only predominate but perhaps also inhibit the PKC-dependent pathway.[72] It is unclear why two distinct mechanisms apparently exist to couple a single receptor to a single functional effect.

Analysis of functional response in endothelial cells also suggests the existence of an $A_{2A}AR$-PKC pathway. Analogous to ischemic preconditioning of cardiac myocytes, it has been shown that exposure of isolated bovine coronary endothelial cells to a brief period of hypoxia significantly protects the cells from injury induced by prolonged ischemia and reoxygenation.[73] The beneficial effects of the preconditioning period could be mimicked by exposure of endothelial cells to an $A_{2A}AR$-selective concentration of CGS21680 prior to ischemia/reoxygenation.[73] Cellular protection afforded by CGS21680 could be completely abolished by pretreatment of cells with the selective PKC inhibitor calphostin C, as was that produced by ischemic preconditioning.[73] The involvement of PKC in the response is supported by the demonstration that phorbol ester treatment prior to ischemia/reoxygenation preserved cell viability to a similar degree as preconditioning and CGS21680.[73] Importantly, as assessed by in vitro phosphorylation of a PKC substrate peptide, treatment of cells with CGS21680 induced a translocation of PKC activity from a cytosolic to a membrane fraction indicative of enzyme activation.[73] In this study, the effect of cAMP analogs or PKA inhibitors on functional or biochemical responses following $A_{2A}AR$ activation was not examined. Activation of the $A_{2A}AR$ has been shown to stimulate adenylyl cyclase in endothelial cells.[74,75] Thus, it would be of interest to determine if the adenylyl cyclase-PKA pathway may have any role in the apparent activation of PKC in response to CGS21680 or if coupling of the $A_{2A}AR$ to PKC is direct. Similarly, it is not known which G protein may couple the $A_{2A}AR$ to PKC stimulation in endothelial cells.

Finally, recent studies have also explored the effect of $A_{2A}AR$ activation on signal transduction via the mitogen-activated protein kinase (MAP kinase) pathway. The studies of Sexl and coworkers[75,76] originated from their characterization of the proliferative effects of adenosine analogs on human umbilical vein endothelial cells (HUVECs). Initially, it was found that activation of a receptor pharmacologically characterized as the $A_{2A}AR$ mediated an increased [³H]thymidine incorporation in these cells.[75] Interestingly, activation of the $A_{2A}AR$ did stimulate adenylyl cyclase activity in HUVECs, however the cell proliferative response was found to occur independently of cAMP accumulation.[75] Furthermore, proliferation in response to NECA was not diminished by prolonged treatment of HUVECs with cholera toxin nor by pretreatment with pertussis toxin suggesting that neither G_s nor G_i coupled the $A_{2A}AR$ to the involved signal transduction pathway.[75] In a subsequent study, this group was able to demonstrate in HUVECs that $A_{2A}AR$ agonists activate the MAP kinase pathway as assessed by the stimulated phosphorylation of the p42 MAP kinase isoform as well as phosphorylation of myelin basic protein.[76] Additionally, NECA induced the translocation of p42 MAP kinase from the periphery of HUVECs to a perinuclear region of the cell indicative of activation.[76] Consistent with the prior characterization of the proliferative response, activation of the MAP kinase pathway by the $A_{2A}AR$ was cAMP-independent and not sensitive to either pertussis or cholera toxin.[76] Finally, stimulation of [³H]thymidine incorporation by activation of the $A_{2A}AR$ in HUVECs was blocked by treatment of the cells with the MAP kinase inhibitor, PD 098059.[76] The possible existence of an $A_{2A}AR$ activated MAP kinase pathway in additional cell types (see below) and the functional consequences of such an activity are areas that need to be explored. The study described above also suggests the interesting possibility that the $A_{2A}AR$ may couple to an as yet unidentified G protein to produce functional response.

$A_{2A}AR$-mediated MAP kinase activation has only been reported for HUVECs, and the importance of cell type in the observation of a particular response is again stressed. Indeed, Hirano and coworkers[77] have demonstrated that activation of the $A_{2A}AR$ in transfected CHO cells results in an ~60% inhibition of the MAP kinase activity stimulated by thrombin in these cells. Based on the results obtained from the combined use of NECA with additional inhibitors of thrombin-stimulated MAP kinase activity, it was proposed that the $A_{2A}AR$-mediated inhibition occurred as a result of adenylyl cyclase stimulation in these cells.[77] However, more direct evidence delineating this pathway is not available.

$A_{2B}AR$

The $A_{2B}AR$ remains the least studied of the four AR subtypes. This is in large part due to the fact that agonist and antagonist ligands that demonstrate selectivity for this receptor are presently unavailable. The lack of selective ligands for the $A_{2B}AR$ not only hinders the study of the biochemical effects elicited by receptor activation but also makes it difficult to unambiguously define physiologic responses as occurring as the result of specifically $A_{2B}AR$ activation. For example, a response is typically categorized as $A_{2B}AR$-mediated if it is elicited by NECA but not the $A_{2A}AR$ selective agonist CGS21680, and displays sensitivity to relatively high concentrations of xanthine antagonists. Obviously, the possibility of multiple ARs existing on the same cell or tissue may make this assignment difficult.

Like the $A_{2A}AR$, the $A_{2B}AR$ couples via G_s to the stimulation of adenylyl cyclase and this response has been shown for the native receptor[78,79] as well as the recombinant receptor expressed in mammalian cells.[80,81] In longitudinal smooth muscle cells isolated from guinea pig intestine, activation of the $A_{2B}AR$ was recently shown to couple positively to adenylyl cyclase as well as induce a relaxation of precontracted cells.[82] It was speculated that activation of PKA following A_{2B}AR stimulation results in an alteration of intracellular calcium levels thus promoting relaxation, though direct evidence is lacking.[82] Likewise, adenosine analogs in a potency order characteristic of the A_{2B}AR have been shown to relax guinea pig aortic rings though the signal transduction mechanism is unknown.[83]

Analogous to the A_{2A}AR, certain studies suggest that the A_{2B}AR may produce functional responses through signal transduction mechanisms not involving adenylyl cyclase stimulation. It has been proposed that activation of the A_{2B}AR, possibly via a cAMP-independent mechanism, may be responsible for the ability of adenosine to potentiate release of allergic mediators from mast cells.[84-86] As described above, a similar functional response has been attributed to A_3AR activation in the rat mast cell model, RBL-2H3[47] and in BDE rats,[54] however some species-dependent variations may exist. Marquardt et al[84] have determined that murine bone marrow-derived mast cells express both the A_{2A}AR and A_{2B}AR as assessed by Northern blotting. The existence of a murine homologue of the A_3AR in these cells was not reported. In the murine mast cells, NECA but not CGS21680 enhanced antigen-induced β-hexosaminidase release indicating involvement of the A_{2B}AR.[84] Subsequently, it was demonstrated that while adenosine substantially increased PKA activity in these cells, the adenosine-induced augmentation of β-hexosaminidase release was not attenuated by the PKA inhibitor, KT5720.[85] Feoktistov and Biaggioni[86] have obtained somewhat similar results in their examination of AR signal transduction in human mast cells. It was found that in the HMC-1 cell line, NECA and CGS21680 both stimulated adenylyl cyclase activity while only NECA was substantially active in promoting the release of interleukin-8 from the cells.[86] Furthermore, NECA but not CGS21680 stimulated accumulation of intracellular inositol phosphates and the apparent release of calcium from intracellular stores in these cells.[86] Interestingly, neither pertussis toxin nor cholera toxin had any effect on the NECA-induced increase in intracellular calcium concentration.[86] It was proposed that in HMC-1 cells that activation of the $A_{2B}AR$ via a non-G_s/G_i protein stimulated phosphoinositide hydrolysis with a resulting elevation in intracellular calcium levels and subsequent mast cell activation.[86] The identification of the $A_{2B}AR$ as the involved AR subtype was based on the finding that responses were elicited by NECA but not CGS21680 (non $A_{2A}AR$) and were blocked by the xanthine AR antagonist, enprofylline (non A_3AR).[86] However, it is not clear if that at the employed concentration enprofylline is ineffective at the human A_3AR.

The ability of the $A_{2B}AR$ to signal via a cAMP-independent mechanism has also been suggested by the ability of the recombinant $A_{2B}AR$ expressed in *Xenopus* oocytes to activate an inward current characteristic of that produced by PLC activation.[87] Additionally, $A_{2B}AR$ signaling in human erythroleukemia cells may also involve a non cAMP-dependent component.[88]

As evident from the above discussion, current studies while carefully performed must unfortunately rely on "negative evidence" to define the role of the $A_{2B}AR$. The further exploration of $A_{2B}AR$ signal transduction, both in regard to molecular mechanisms and physiologic significance, would be greatly aided by the development of subtype selective ligands.

STRUCTURAL DETERMINANTS OF AR SIGNAL TRANSDUCTION

In that several cells such as cardiac myocytes contain multiple ARs, the ultimate response of the cell upon either accumulation of endogenously-released adenosine or the administration of its analogs, is dictated predominantly by the relative affinity of the specific AR subtypes for the agonist, as well as the selective coupling of a particular receptor subtype to a specific G protein and that G protein to effector. Following agonist binding by the receptor and the associated conformational change in receptor structure, the initial step in signal transduction is the activation of the heterotrimeric G protein complex. In the cellular environment, the selective interaction of receptor and G protein and thus fidelity of signal transduction may be regulated by several factors including the relative affinities of receptor for the complement of G proteins, the relative amounts of individual G proteins present and perhaps the localization of receptor and G protein in distinct compartments at the cell membrane. The affinity of activated receptor for various G proteins is in large part dictated by the physical features of both the receptor as well as the G protein. In regard to the structural features of the G protein complex that directly interact with receptor, focus has predominantly been on the G protein α subunit,[89,90] however, more recent studies have also examined the G protein βγ complex (reviewed in ref. 91).

Molecular models of receptor-G protein interaction that for the most part are derived from the crystal structure of G protein α subunits and results from mutagenesis studies of both receptors and G proteins have recently been reviewed.[92,93] Much of the available data regarding key structural features of the receptor that are involved in G protein interaction are derived from extensive studies of rhodopsin and the adrenergic and muscarinic receptors.[94-96] It is expected that several of the features of these models are applicable to the vast majority of G protein coupled receptors. For ARs in particular, there are currently no data available to suggest otherwise. However, it is noted that the precise conformational change in receptor structure required for productive G protein coupling is not completely defined for any receptor. Differences exist between ARs and those receptors that are activated by catecholamines in regard to regions involved in ligand recognition.[97] The degree to which these differences may account for significant variations in the mechanism by which the individual receptors transduce the activation signal to the G protein located at the cytosplasmic face of the cell membrane is not known with certainty. However, based on the fact that receptor regions involved in G protein activation are quite similar for receptors with remarkably varied ligand binding properties, it has been proposed that such differences may be minimal.[93] Provided below are the findings of studies that have examined the basic mechanisms and structural features involved in specifically AR-G protein interaction. Focus will be initially placed on findings regarding the selectivity of G protein interactions with ARs.

Freissmuth and coworkers[98,99] have extensively examined the selectivity of G protein interaction with the A_1AR. In an initial study employing preparations of the A_1AR purified from bovine brain mixed in detergent solution with purified bovine brain G protein $\beta\gamma$ complex and various recombinant $G_{\alpha i}$ subunits isolated from *E. coli* lysates, it was found that all three isoforms of $G_{\alpha i}$ as well as $G_{\alpha o}$ could functionally interact with the A_1AR as assessed by restoration of high affinity agonist binding by the receptor.[98] As might be expected, the receptor high affinity state was not achieved following recombination of the A_1AR with the long and short isoforms of $G_{\alpha s}$ or $G_{\alpha z}$.[98] Interestingly, titration of the A_1AR with increasing amounts of $G_{\alpha i}$ subunits revealed a 10-fold greater affinity of the receptor for specifically $G_{\alpha i-3}$ compared to the other $G_{\alpha i}$ isoforms.[98]

In a subsequent study employing either recombinant human A_1AR produced by *E. coli* or human brain membranes reconstituted with purified G protein α subunits, it was found that the human A_1AR, unlike the bovine homologue, interacted equally well with all three $G_{\alpha i}$ isoforms.[99] In these reconstituted systems, discrimination by a particular receptor for a distinct G protein α subunit would be expected to arise from predominantly structural differences as factors that may regulate in vivo coupling, such as G protein stoichiometry and membrane compartmentalization, have been controlled for or eliminated. The apparent species dependence in selectivity of A_1AR-G protein coupling is somewhat surprising based on the high degree of sequence identity between these receptors, however, very limited differences in amino acid composition have been shown to account for marked variations in ligand binding by the canine and bovine A_1ARs.[100] The amino acids responsible for this differential G protein coupling have not been identified. If amino acid differences account for the apparent species selectivity in G protein coupling, the involved receptor regions could directly interact with the G protein or perhaps influence the conformation of critical intracellular domains. In vivo, additional proteins may also influence A_1AR-G protein coupling. Nanoff et al[101] have characterized a "coupling factor" that is responsible for the well documented relative insensitivity of the A_1AR-G protein interaction to guanine nucleotide treatment. Interestingly, the activity of the coupling factor was apparent in rat and bovine brain membranes but not in human brain membranes which may reflect a low affinity interaction between the factor and the human A_1AR.[101] Though partially purified, the identity of this protein has not been determined.

The importance of G protein $\beta\gamma$ subunit composition in coupling of activated A_1AR to the G protein heterotrimeric complex has been examined by Figler and coworkers.[102] This study employed membranes obtained from Sf9 cells expressing recombinant bovine A_1AR reconstituted with recombinant G protein α and $\beta\gamma$ subunits purified from baculovirus infected Sf9 cells. The A_1AR interacted very poorly with the native G protein complexes of Sf9 cells as evidenced by a low degree of high affinity binding in the absence of added G protein α and $\beta\gamma$ subunits. However, upon addition of G protein α subunits and $\beta\gamma$ complex of defined composition, high affinity guanine nucleotide sensitive agonist binding was demonstrated by the A_1AR. It was determined that when added with various $G_{\alpha i}$ isoforms, five different $\beta\gamma$ subunit combinations were equally efficacious in promoting high affinity binding by the A_1AR. However, a complex of specifically $G_{\beta 1\gamma 1}$ was less effective in promoting A_1AR-G protein coupling. It should be noted that in this study of bovine A_1AR-G protein coupling that the receptor did not discriminate between $G_{\alpha i}$ subunits in reconstitution assays in contrast to the findings described previously. One possible explanation is the expression system employed to supply purified G protein α subunits, i.e., *E. coli* vs. Sf9 cells. It is possible that post-translational modifications of G proteins required for efficient interaction with receptor may vary between bacteria and insect cells. Indeed, the specific nature of the prenyl group on the carboxyl terminus of the G protein γ subunit has recently been shown to be an important determinant of productive coupling of recombinant bovine A_1AR with G protein α subunits and $\beta\gamma$ dimers in reconstituted systems.[103]

The above discussion focused on the data currently available regarding A_1AR interactions with G protein complexes in various in vitro conditions. In the intact cell, specificity of receptor-G protein coupling is subject to control by a multitude of additional factors as discussed previously. The specificity of A_1AR coupling to specific G proteins has not been extensively examined in vivo. As noted above, a recent study employing rabbit intestinal smooth muscle demonstrated that A_1AR-mediated adenylyl cyclase inhibition was sensitive specifically to antisera raised against $G_{\alpha i-3}$ despite the presence of additional $G_{\alpha i}$ isoforms in the cell.[17] In that reconstitution studies indicate that the A_1AR can efficiently couple to all isoforms of $G_{\alpha i}$, an obvious question regards the cellular factors that apparently dictate specificity in vivo.

Very little is known regarding the specificity of A_3AR coupling to various G proteins. It has been shown in CHO cells heterologously expressing the rat A_3AR that stimulation of the receptor by agonist results in activation of both $G_{\alpha i-2}$ and $G_{\alpha i-3}$ to approximately a similar extent as assessed by specific incorporation of the G protein photoaffinity probe, 4-azidoanilido-$[\alpha-^{32}P]$-GTP (^{32}P-AA-GTP).[104] Interestingly, agonist-stimulated ^{32}P-AA-GTP labeling also indicated that the A_3AR weakly interacted with α subunits of the $G_{q/11}$ family, though currently there are no functional data that suggest such a coupling.[104] Similarly, it is not known in any system which specific $G_{\alpha i}$ or $G_{\beta\gamma}$ subunits are involved in mediating signal transduction initiated by A_3AR activation.

In regard to the structural features of ARs that are responsible for selective coupling to G proteins, there are no published reports that have examined the A_1AR, A_3AR or $A_{2B}AR$. It is interesting to note that the cloning of a splice variant of the rat A_3AR has recently been reported.[105] Relative to the initially isolated rat A_3AR, this receptor termed A3i by Sajjadi and coworkers, contains a 17 amino acid insert in the mid-portion of the second intracellular loop.[105] In that this cytoplasmic region may be involved in G protein coupling, such a variation may influence receptor signal transduction. In the characterization of A3i, it was found that following stable transfection into CHO cells, the receptor mediated minimal inhibition of adenylyl cyclase activity suggesting poor coupling to G_i.[105] Based on rather low specific radioligand binding displayed by this cell line, it is possible that the relative lack of functional effect may occur simply due to poor receptor expression as agonist efficacy is often a function of receptor expression. Conversely, the low specific binding by the agonist ^{125}I-ABA may also reflect a large receptor population unable to form high affinity complex with the G protein complement present in CHO cells. The availability of an antagonist radioligand for the rat A_3AR would be of benefit to further study function and expression of this receptor.

The structural determinants of the $A_{2A}AR$ that are involved in selective coupling to G_s have recently been examined in a mutational analysis of canine $A_{2A}AR$-mediated adenylyl cyclase stimulation.[106] Chimeric ARs in which distinct intracellular domains of the $A_{2A}AR$ were replaced with the analogous segments of the G_i-coupled bovine A_1AR, were created and transiently transfected into CHO cells, and adenylyl cyclase stimulation in response to NECA was determined. Replacement of the entire third intracellular loop of the $A_{2A}AR$ with that of the A_1AR resulted in a receptor that in response to NECA demonstrated a maximal stimulation of adenylyl cyclase that was ~25% that of the wild-type $A_{2A}AR$ and agonist potency was decreased five-fold. Conversely, replacement of the 114 amino acid COOH-terminal tail of the $A_{2A}AR$ with the 36 amino acid tail of the A_1AR had no effect on the ability of NECA to stimulate adenylyl cyclase. More defined mutagenesis of the third intracellular loop of the $A_{2A}AR$ indicated that coupling of the receptor to G_s was predominantly dictated by a 15 amino acid cassette located at the most NH_2-terminal portion of this loop. Point mutations in this region indicated that lysine-209 and glutamic acid-212 were involved in the efficiency of $A_{2A}AR$ coupling to G_s as assessed by the potency of NECA in adenylyl cyclase assays. The importance of the identified 15 amino acid segment of the $A_{2A}AR$ in G_s activation was further verified by creation of the reciprocal chimeric receptor in which the 15 amino acids of the NH_2-terminal portion of the third intracellular loop of the $A_{2A}AR$ replaced those of the A_1AR. Upon expression in CHO cells, this chimera produced negligible stimulation of adenylyl cyclase. However, when cells were pretreated with pertussis toxin to ablate any residual coupling of this construct composed predominantly of A_1AR structure to G_i, this chimeric receptor demonstrated a significant stimulation of adenylyl cyclase that was ~25% of that induced by the wild-type $A_{2A}AR$. Membranes from pertussis toxin treated cells expressing the wild-type A_1AR displayed no stimulation of adenylyl cyclase. Thus, the 15 amino acids derived from the $A_{2A}AR$ were sufficient to promote a degree of G_s coupling to the chimeric receptor. The chimeric receptor displayed a potency for agonist identical to that of the wild-type A_1AR for inhibition of adenylyl cyclase indicative of efficient coupling. Interestingly, a chimeric receptor in which the entire third intracellular loop of the A_1AR was replaced with that of the $A_{2A}AR$ stimulated adenylyl cyclase to a similar degree in the presence and absence of pertussis toxin.

A similar approach of creating chimeric $A_1/A_{2A}ARs$ was taken to examine the role of the second intracellular loop of the $A_{2A}AR$ in G_s coupling. It was found that the nature of the amino acids constituting the junction between the second intracellular loop and transmembrane domain 4 of the receptor is important in maintaining efficient G protein coupling. When these residues of the $A_{2A}AR$ (glycine and threonine) were replaced with the analogous residues of the A_1AR (proline and arginine), the potency of NECA in adenlylyl cyclase assays decreased ~50-fold relative to wild-type $A_{2A}AR$. Additionally, agonist competition binding assays also reflected altered receptor-G_s coupling by this tandem amino acid substitution. However, these residues may not directly interact with the G protein but perhaps alter the conformation of other directly acting regions. The individual replacement of either the glycine or threonine did not affect coupling nor did the dual replacement of these amino acids with alanines.

To summarize this study of $A_{2A}AR$-G_s coupling, it appears that the predominant structural feature of the receptor dictating selective G_s interaction is the NH_2-terminal region of the third intracellular loop. This is consistent with other studies of G protein coupled receptors.[95,96] The profile of reciprocal chimeras strongly supports this conclusion, though as with all mutagenesis studies of this nature, it is not possible to unambiguously define the direct versus indirect effects of amino acid replacement. It would be of interest to further define the minimal amount of amino acid sequence necessary for productive receptor-G_s coupling. Additionally, as described previously, there is an increasing amount of data that suggest that the $A_{2A}AR$ does not exclusively couple to G_s to produce cellular effects. It would be of interest to perform similar mutational analysis of the $A_{2A}AR$, perhaps monitoring protein kinase C activation or another functional response. This type of study would require a well-defined and somewhat robust readout response which would necessitate the identification of a suitable cell system to employ. In regard to $A_{2A}AR$ signal transduction, it is also interesting to note that the cytoplasmic tail of the receptor does not appear to influence the selective coupling to G_s.[106] It has also been demonstrated that the majority of this tail may be deleted without influencing $A_{2A}AR$ agonist-induced desensitization of adenylyl cyclase activity in CHO cells.[107] In that this cytoplasmic region of the $A_{2A}AR$ is approximately 80 amino acids larger than that of any other AR subtype, a remaining question regards the function of this domain. Finally, similar mutagenesis studies have not been published for any of the other AR subtypes. In addition to identifying regions of the A_1AR and A_3AR involved in G_i coupling, it would be of interest to further determine why the A_1AR but not the A_3AR appears to couple more efficiently to adenylyl cyclase inhibition. It is possible that this may be due to structural features of the receptors or simply differences in the G proteins and/or effector molecules in cells which predominantly express the A_1AR and A_3AR. A more thorough examination of the ability of the A_3AR to inhibit adenylyl cyclase activity in native cells is warranted.

CONCLUSIONS

Adenosine can elicit responses in nearly all organ systems. Provided in this chapter has been a review of the signal transduction mechanisms employed by ARs to produce these effects. Considering the many systems responsive to adenosine, the clinical regulation of ARs has been rather limited. Based on the current widespread utilization of drugs acting on various G protein coupled receptors, it is speculated that ARs may also serve as therapeutic targets. Further exploration of the basic mechanisms of AR activity may help advance the more extensive clinical manipulation of adenosine signaling pathways. In various sections of this review, possible directions for future research have been suggested. These include the extension of findings regarding basic mechanisms of signal transduction from in vitro systems to the more physiologic setting. Investigation into cell-specific effects of AR activation should continue as signal transduction paradigms may be dependent on the cell type under examination. Additionally, advances need to be made in the development of potent and subtype selective agonists and antagonists particularly for the A_3AR and $A_{2B}AR$. Finally, exploration of possible alterations in AR function or regulation in specific pathophysiologic conditions may provide the basis for the potential for therapeutic manipulation of AR signaling to be realized.

REFERENCES

1. Olah ME, Stiles, GL. Adenosine receptor subtypes: characterization and therapeutic regulation. Annu Rev Pharmacol Toxicol 1995; 35:581-606.
2. Tucker AL, Linden J. Cloned receptors and cardiovascular responses to adenosine. Cardiovasc Res 1993; 27:62-67.
3. Olsson RA, Pearson JD. Cardiovascular purinoceptors. Physiol Rev 1990; 70:761-845.
4. Belardinelli L, Shryock JC, Pelleg A. Cardiac electrophysiologic properties of adenosine. Coronary Artery Disease 1992; 3:1122-1126.
5. Mubagwa K, Mullane K, Flameng W. Role of adenosine in the heart and circulation. Cardiovasc Res 1996; 32:797-813.
6. Latini S, Pazzagli M, Pepeu G et al. A_2 adenosine receptors: their presence and neuromodulatory role in the central nervous system. Gen Pharmac 1996; 27:925-933.

7. Van Calker D, Muller M, Hamprecht B. Adenosine regulates two different types of receptors, the accumulation of cyclic AMP in cultured brain cells. J Neurochem 1979; 33:999-1005.

8. Londos C, Cooper DMF, Wolff J. Subclasses of external adenosine receptors. Proc Nat Acad Sci USA 1980; 77:2551-2554.

9. Ebersolt C, Premonst J, Prochiantz A et al. Inhibition of brain adenylate cyclase by A1 adenosine receptors: Pharmacological characteristics and locations. Brain Res 1983; 267:123-129.

10. Ma H, Green RD. Modulation of cardiac cyclic AMP metabolism by adenosine receptor agonists and antagonists. Mol Pharmacol 1992; 42:831-837.

11. Ramkumar V, Barrington WW, Jacobson KA et al. Demonstration of both A_1 and A_2 adenosine receptors in DDT_1 MF-2 smooth muscle cells. Mol Pharmacol 1989; 37:149-156.

12. Delahunty, TM, Cronin, MJ, Linden, J. Regulation of GH_3-cell function via adenosine A_1 receptors: Inhibition of prolactin relaease, cyclic AMP production and inositol phosphate generation. Biochem J 1988; 255:69-77.

13. Mahan LC, McVittie LD, Smyk-Randall EM et al. Cloning and expression of an A_1 adenosine receptor from rat brain. Mol Pharmacol 1991; 40:1-7.

14. Olah ME, Ren H, Ostrowski J et al. Cloning, expression and characterization of the unique bovine A_1 adenosine receptor: studies on the ligand binding site by site-directed mutagenesis. J Biol Chem 1992; 267:10764-10770.

15. Taussig R, Gilman AG. Mammalian membrane-bound adenylyl cyclases. J Biol Chem 1995; 270:1-4.

16. Ishikawa Y, Homcy CJ. The adenylyl cyclases as integrators of transmembrane signal transduction. Circ Res 1997; 80:297-304.

17. Murthy KS, Makhlouf GM. Adenosine A_1 receptor-mediated activation of phospholipase C-β_3 in intestinal muscle: Dual requirement for α and $\beta\gamma$ subunits of G_{i3}. Mol Pharmacol 1995; 47:1172-1179.

18. Belardinelli L, Vogel S, Linden J et al. Antiadrenergic action of adenosine on ventricular myocardium in embryonic chick hearts. J Mol Cell Cardiol 1982; 14:291-294.

19. Dobson JG, Schrader J. Role of extracellular and intracellular adenosine in the attenuation of catecholamine evoked responses in guinea pig heart. J Mol Cell Cardiol 1984; 16:813-822.

20. Clapham DE, Neer J. G protein $\beta\gamma$ subunits. Annu Rev Pharmacol Toxicol 1997; 37:167-203.

21. Logothetis DE, Kurachi Y, Galper J et al. The $\beta\gamma$ subunits of GTP-binding proteins activate the muscarinic K^+ channel in heart. Nature 1987; 325:321-326.

22. Kurachi Y, Nakajima T, Sugimoto T. On the mechanism of activation of muscarinic K$^+$ channels by adenosine in isolated atrial cells: Involvement of GTP-binding proteins. Pflugers Arch 1986; 407: 264-274.

23. Arend LJ, Handler JS, Rhim JS et al. Adenosine-sensitive phosphoinositide turnover in a newly established renal cell line. Am J Physiol 1989; 256:F1067-F1074.

24. Gerwins P, Fredholm BB. ATP and its metabolite adenosine act synergistically to mobilize intracellular calcium via the formation of inositol 1,4,5-trisphosphate in a smooth muscle cell line. J Biol Chem 1992; 267:16081-16087.

25. Peakman M-C, Hill SJ. Adenosine A$_1$ receptor-mediated changes in basal and histamine-stimulated levels of intracellular calcium in primary rat astrocytes. Br J Pharmacol 1995; 115:801-810.

26. Freund S, Ungerer M, Lohse MJ. A$_1$ adenosine receptors expressed in CHO-cells couple to adenylyl cyclase and to phospholipase C. Naunyn-Schmiedeberg's Arch Pharmacol 1994; 350:49-56.

27. Iredale PA, Alexander SPH, Hill SJ. Coupling of a transfected human brain A$_1$ adenosine receptor in CHO-K1 cells to calcium mobilisation via a pertussis toxin-sensitive mechanism. Br J Pharmacol 1994; 111:1252-1256.

28. Akbar M, Okajima F, Tomura H et al. A single species of A$_1$ adenosine receptor expressed in Chinese hamster ovary cells not only inhibits cAMP accumulation but also stimulates phospholipase C and arachidonate release. Mol Pharmacol 1994; 45:1036-1042.

29. Megson AC, Dickenson JM, Townsend-Nicholson A et al. Synergy between the inositol phosphate responses to transfected human adenosine A$_1$-receptors and constitutive P$_2$-purinoceptors in CHO-K1 cells. Br J Pharmacol 1995; 115:1415-1424.

30. Koch WJ, Hawes BE, Inglese J et al. Cellular expression of the carboxyl terminus of a G-protein coupled receptor kinase attenuates βγ-mediated signaling. J Biol Chem 1994; 269:6193-6197.

31. Smrcka AV, Sternweis PC. Regulation of purified subtypes of phosphatidylinositol-specific phospholipase C β by G protein α and βγ subunits. J Biol Chem 1993; 267:9667-9674.

32. Cohen MV, Downey JM. Myocardial preconditioning promises to be a novel approach to the treatment of ischemic heart disease. Annu Rev Med 1996; 47:21-29.

33. Liu Y, Gao WD, O'Rourke B et al. Synergistic modulation of ATP-sensitive K$^+$ currents by protein kinase C and adenosine. Circ Res 1996; 78:443-454.

34. Yao Z, Mizumura T, Mei DA et al. K$_{ATP}$ channels and memory of ischemic preconditioning in dogs: synergism between adenosine and K$_{ATP}$ channels. Am J Physiol 1997; 272:H334-H342.

35. Sakamoto J, Miura T, Goto M et al. Limitation of myocardial infarct size by adenosine A_1 receptor activation is abolished by protein kinase C inhibitors in the rabbit. Cardiovasc Res 1995; 29:682-688.

36. Henry P, Demolombe S, Pucéat M et al. Adenosine A_1 stimulation activates δ-protein kinase C in rat ventricular myocytes. Circ Res 1996;78:161-165.

37. Nanoff C, Freissmuth M, Tuisl E et al. P_2-, but not P_1-purinoceptors mediate formation of 1,4,5-inositol trisphosphate and its metabolites via a pertussis toxin-insensitive pathway in the rat renal cortex. Br J Pharmacol 1990; 100:63-68.

38. Long CJ, Stone TW. Adenosine reduces agonist-induced production of inositol phosphates in rat aorta. J Pharm Pharmacol 1987; 39:1010-1014.

39. Zhou Q-Y, Li C, Olah ME et al. Molecular cloning and characterization of an adenosine receptor: The A_3 adenosine receptor. Proc Natl Acad Sci USA 1992; 89:7432-74336.

40. Linden J, Taylor HE, Robeva AS et al. Molecular cloning and functional expression of a sheep A_3 adenosine receptor with widespread tissue distribution. Mol Pharmacol 1993; 44:524-532.

41. Salvatore CA, Jacobson MA, Taylor HE et al. Molecular cloning and characterization of the human A_3 adenosine receptor. Proc Natl Acad USA 1993; 90:10365-10369.

42. Jacobson KA, Nikodijevic O, Shi D et al. A role for central A_3 adenosine receptors: mediation of behavioral depressant effects. FEBS Lett 1993; 336:57-60.

43. Abbracchio MP, Brambilla R, Ceruti S et al. G protein-dependent activation of phospholipase C by adenosine A_3 receptors in rat brain. Mol Pharmacol 1995; 48:1038-1045.

44. Jacobson KA, Park KS, Jiang J-I et al. Pharmacological characterization of novel A_3 adenosine receptor-selective antagonists. Neuropharmacology 1997; 36:1157-1165.

45. Strickler J, Jacobson KA, Liang BT. Direct preconditioning of cultured chick ventricular myocytes. Novel functions of cardiac adenosine A_{2a} and A_3 receptors. J Clin Invest 1996; 98:1773-1779.

46. Ali H, Cunha-Melo JR, Saul WF et al. Activation of phospholipase C via adenosine receptors provides synergistic signals for secretion in antigen-stimulated RBL-2H3 cells: Evidence for a novel adenosine receptor. J Biol Chem 1990; 265:745-753.

47. Ramkumar V, Stiles GL, Beaven MA et al. The A_3 adenosine receptor is the unique adenosine receptor which facilitates release of allergic mediators in mast cells. J Biol Chem 1993; 268:16887-16890.

48. Antonysamy MA, Moticka EJ, Ramkumar V. Adenosine acts as an endogenous modulator of IL-2-dependent proliferation of cytotoxic T lymphocytes. J Immunol 1995; 155:2813-2821.

49. Kohno Y, Sei Y, Koshiba M et al. Induction of apoptosis in HL-60 human promyelocytic leukemia cells by adenosine A_3 receptor agonists. Biochem Biophys Res Commun 1996; 219:904-910.

50. Kohno Y, Ji X-d, Mawhorter SD et al. Activation of A_3 adenosine receptors on human eosinophils elevates intracellular calcium. Blood 1996; 88:3569-3574.

51. Beaven JA, Ramkumar V, Ali H. Adenosine A_3 receptors in mast cells. Trends Pharmacol Sci 1994; 15:13-14.

52. Hannon JP, Pfannkuche H-J, Fozard JR. A role for mast cells in adenosine A_3 receptor-mediated hypotension in the rat. Br J Pharmacol 1995; 115:945-952.

53. Shepherd RK, Linden J, Duling BR. Adenosine-induced vasoconstriction in vivo. Role of the mast cell and A_3 adenosine receptor. Circ Res 1996; 78-627-634.

54. Meade CJ, Mierau J, Leon I et al. In vivo role of the adenosine A_3 receptor: N^6-2-(4-aminophenyl)ethyladenosine induces bronchospasm in BDE rats by a neurally mediated mechanism involving cells resembling mast cells. J Pharmacol Exp Ther 1996; 279:1148-1156.

55. Armstrong S, Ganote CE. Adenosine receptor specificity in preconditioning of isolated rabbit cardiomyocytes: Evidence of A_3 receptor involvement. Cardiovasc Res 1994; 28:1049-1056.

56. Auchampach JA, Rizvi A, Qiu Y et al. Selective activation of A_3 adenosine receptors with N^6-(3-Iodobenzyl)adenosine-5'-N-methyluronamide protects against myocardial stunning and infarction without hemodynamic changes in conscious rabbits. Circ Res 1997; 80:800-809.

57. Omitted in proof.

58. Cooper JA, Hill SJ, Alexander SPH et al. Adenosine receptor-induced cyclic AMP generation and inhibition of 5-hydroxytryptamine release in human platelets. Br J Clin Pharmacol 1995; 40:43-50.

59. Fredholm BB, Zhang Y, van der Ploeg I. Adenosine A_{2A} receptors mediate the inhibitory effect of adenosine on formyl-Met-Leu-Phe-stimulated respiratory burst in neutrophil leucocytes. Naunyn-Schmiedeberg's Arch Pharmacol 1996; 354:262-267.

60. Walker BAM, Rocchini C, Boone RH et al. Adenosine A_{2a} receptor activation delays apoptosis in human neutrophils. J Immunol 1997; 158:2926-2931.

61. Cronstein BN, Haines KA, Kolasinski S et al. Occupancy of $G_{\alpha s}$-linked receptors uncouples chemotractant receptors from their stimulus transduction mechanisms in the neutrophil. Blood 1992; 80: 1052-1057.

62. Cushing DJ, Brown GL, Sabouni MH et al. Adenosine receptor-mediated coronary artery relaxation and cyclic nucleotide production. Am J Physiol 1991; 261:H343-H348.

63. Kleppisch T, Nelson MT. Adenosine activates ATP-sensitive potassium channels in arterial myocytes via A_2 receptors and cAMP-dependent protein kinase. Proc Natl Acad Sci USA 1995; 92: 12441-12445.

64. Jackson WF. Arteriolar tone is determined by activity of ATP-sensitive potassium channels. Am J Physiol 1993; 265:H1797-H1803.

65. Abebe W, Hussain T, Olanreqaju H et al. Role of nitric oxide in adenosine receptor-mediated relaxation of porcine coronary artery. Am J Physiol 1995; 269:H1672-H1678.

66. Liang BT, Morley JF. A new cyclic AMP-independent, G_s-mediated stimulatory mechanism via the adenosine A_{2a} receptor in the intact cardiac cell. J Biol Chem 1995; 271:18678-18685.

67. Yatani Y, Brown AM. Rapid β-adrenergic modulation of cardiac calcium channel currents by a fast G protein pathway. Science 1989; 245:71-74.

68. Hide I, Padgett WL, Jacobson KA et al. A_{2A} adenosine receptors from rat striatum and rat pheochromocytoma PC12 cells: Characterization with radioligand binding and by activation of adenylate cyclase. Mol Pharmacol 1992; 41:352-359.

69. Chern Y, Lai H-L, Fong JC et al. Multiple mechanisms for desensitization of A_{2a} adenosine receptor-mediated cAMP elevation in rat pheochromocytoma PC12 cells. Mol Pharmacol 1993; 44:950-958.

70. Koizumi S, Watano T, Nakazawa K et al. Potentiation by adenosine of ATP-evoked dopamine release via a pertussis toxin-sensitive mechanism in rat phaeochromocytoma PC12 cells. Br J Pharmacol 1994; 112:992-997.

71. Lai H-L, Yang T-H, Messing RO et al. Protein kinase C inhibits adenylyl cyclase type VI activity during desensitization of the A_{2a}-adenosine receptor-mediated cAMP response. J Biol Chem 1997; 272:4970-4977.

72. Gubitz AK, Widdowson L, Kurokawa M et al. Dual signalling by the adenosine A_{2a} receptor involves activation of both N- and P-type calcium channels by different G proteins and protein kinases in the same striatal nerve terminals. J Neurochem 1996; 67:374-381.

73. Zhou X, Zhai X, Ashraf M. Preconditioning of bovine endothelial cells. The protective effect is mediated by an adenosine A_2 receptor through a protein kinase C signaling pathway. Circ Res 1996; 78:73-81.

74. Iwamoto T, Umemura S, Toya Y et al. Identification of adenosine A_2 receptor-cAMP system in human aortic endothelial cells. Biochem Biophys Res Commun 1994; 199:905-910.

75. Sexl V, Mancusi G, Baumgartner-Parzer S et al. Stimulation of human umbilical vein endothelial cell proliferation by A_2-adenosine and $β_2$-adrenoceptors. Br J Pharmacol 1995; 114:1577-1586.

76. Sexl V, Mancusi G, Höller C et al. Stimulation of the mitogen-activated protein kinase via the A_{2A}-adenosine receptor in primary human endothelial cells. J Biol Chem 1997; 272:5792-5799.

77. Hirano D, Aoki Y, Ogasawara H et al. Functional coupling of adenosine A_{2a} receptor to inhibition of the mitogen-activated protein kinase cascade in Chinese hamster ovary cells. Biochem J 1996; 316:81-86.

78. Lupica CR, Cass WA, Zahniser NR et al. Effects of the selective adenosine A2 receptor agonist CGS21680 on in vitro electrophysiology, cAMP formation and dopamine release in rat hippocampus and striatum. J Pharmacol Exp Ther 1990; 252:1134-1141.

79. Brackett LE, Daly JW. Functional characterization of the A_{2b} adenosine receptor in NIH 3T3 fibroblasts. Biochem Pharmacol 1994; 47:801-814.

80. Rivkees SA, Reppert SM. RFL9 encodes an A_{2b}-adenosine receptor. Mol Endocrinol 6:1598-1604.

81. Pierce KD, Furlong TJ, Selbie LA et al. Molecular cloning and expression of an adenosine A_{2b} receptor from human brain. Biochem Biophys Res Commun 1992; 187:86-93.

82. Murthy KS, McHenry L, Grider JR et al. Adenosine A_1 and A_{2b} receptors coupled to distinct interactive signaling pathways in intestinal muscle cells. J Pharmacol Exp Ther 1995; 274:300-306.

83. Gurden MF, Coates J, Ellis F et al. Function characterization of three adenosine receptor types. Br J Pharmacol 1993; 109:693-698.

84. Marquardt DL, Walker LL, Heinemann S. Cloning of two adenosine receptor subtypes from mouse bone marrow-derived mast cells. J Immunol 1994; 152:4508-4515.

85. Marquardt DL, Walker LL. Inhibition of protein kinase A fails to alter mast cell adenosine responsiveness. Agents Actions 1994; 43:7-12.

86. Feoktistov I, Biaggioni I. Adenosine A_{2b} receptors evoke interleukin-8 secretion in human mast cells. J Clin Invest 1995; 96:1979-1986.

87. Yakel JL, Warren RA, Reppert SM et al. Functional expression of adenosine A_{2b} receptor in *Xenopus* oocytes. Mol Pharmacol 1993; 43:277-280.

88. Feoktistov I, Murray JJ, Biaggioni I. Positive modulation of intracellular Ca^{2+} levels by adenosine A_{2b} receptors, prostacyclin, and prostaglandin E_1 via a cholera toxin-sensitive mechanism in human erythroleukemia cells. Mol Pharmacol 1994; 45:1160-1167.

89. Conklin BR, Farfel Z, Lustig KD et al. Substitution of three amino acids switches receptor specificity of $G_q\alpha$ to that of $G_i\alpha$. Nature 1993; 363:274-276.

90. Rens-Domiano S, Hamm HE. Structural and functional relationships of heterotrimeric G-proteins. FASEB J 1995; 9:1059-1066.

91. Gudermann T, Kalkbrenner F, Schultz G. Diversity and selectivity of receptor-G protein interaction. Annu Rev Pharmacol Toxicol 1996; 36:429-459.

92. Baldwin JM. Structure and function of receptors coupled to G proteins. Cur Op Cell Biol 1994; 6:180-190.

93. Bourne HR. How receptors talk to trimeric G proteins. Curr Op Cell Biol 1997; 9:134-142.
94. Franke RR, König B, Sakmar TP et al. Rhodopsin mutants that bind but fail to activate transducin. Science 1990; 250:123-125.
95. Strader CD, Fong TM, Tota MR et al. Structure and function of G protein-coupled receptors. Annu Rev Biochem 1994; 63:101-132.
96. Wess J, Blin N, Mutschler E et al. Muscarinic acetylcholine receptors: Structural basis of ligand binding and G protein coupling. Life Sci 1995; 56:915-922.
97. Olah ME, Jacobson KA, Stiles GL. Role of the second extracellular loop of adenosine receptors in agonist and antagonist binding: Analysis of chimeric A_1/A_3 receptors. J Biol Chem 1994; 269:24692-24698.
98. Freissmuth M, Schütz W, Linder M. Interaction of the bovine brain A_1-adenosine receptor with recombinant G protein α-subunits. J Biol Chem 1991; 266:17778-17783.
99. Jockers R, Linder ME, Hohenegger M et al. Species differences in the G protein selectivity of the human and the bovine A1 adenosine receptor. J Biol Chem 1994; 269:32077-32084.
100. Tucker AL, Robeva AS, Taylor HE et al. A_1 adenosine receptors: two amino acids are responsible for species differences in ligand recognition. J Biol Chem 269:27900-27906.
101. Nanoff C, Mitterauer T, Roka F et al. Species differences in A_1 adenosine receptor/G protein coupling: Identification of a membrane protein that stabilizes the association of the receptor/G protein complex. Mol Pharmacol 1995; 48:806-817.
102. Figler RA, Graber SG, Lindorfer MA et al. Reconstitution of recombinant bovine A_1 adenosine receptors in Sf9 cell membranes with recombinant G proteins of defined composition. Mol Pharmacol 1996; 50:1587-1595.
103. Yasuda H, Lindorfer MA, Woodfork KA et al. Role of the prenyl group on the G protein γ subunit in coupling trimeric G proteins to A1 adenosine receptors. J Biol Chem 1996; 271:18588-18595.
104. Palmer TM, Gettys TW, Stiles GL. Differential interaction with and regulation of multiple G-proteins by the rat A3 adenosine receptor. J Biol Chem 1995; 270:16895-16902.
105. Sajjadi FG, Boyle DL, Domingo RC et al. cDNA cloning and characterization of A3i, an alternatively spliced rat A3 adenosine receptor variant. FEBS Letters 1996; 382:125-129.
106. Olah ME. Identification of A_{2a} adenosine receptor domains involved in selective coupling to G_s. J Biol Chem 1997; 272:337-344.
107. Palmer TM, Stiles GL. Identification of an A_{2a} adenosine receptor domain specifically responsible for mediating short-term desensitization. Biochem 1997; 36:832-838.

CHAPTER 2

Adenosine Metabolism and Transport in the Mammalian Heart[#]

Jürgen Schrader, Andreas Deussen and Ulrich K.M. Decking

Continuous interest in the role of adenosine results from the multiple pharmacological and physiological actions of this nucleoside in various cell types and organs. It was as early as 1929 when Drury and Szent Gyorgyi first described the cardiac actions of adenosine.[39] These include potent coronary vasodilation, reversible slowing the heart rate, impairment of atrioventricular conduction and the antiarrhythmic effect of adenosine. Two years later Lindner and Rigler[69] advanced the hypothesis that adenosine is a physiological regulator of coronary blood flow. In the sixties it was recognized by Berne[15] and Gerlach's group[43] that the heart in fact can produce adenosine during hypoxic perfusion and ischemia. This observation has triggered a multitude of studies aimed to

 # This review is dedicated to Keith Kroll, Ph.D., who was asked to write this chapter, but was unable to do so due to his early death (December 9, 1948 - July 16, 1997). Keith has substantially contributed to our present understanding of the metabolism of adenosine and its function. We have lost a dear friend and an outstanding scientist.

Effects of Extracellular Adenosine and ATP on Cardiomyocytes,
edited by Amir Pelleg and Luiz Belardinelli. © 1998 R.G. Landes Company.

elucidate the functional role of adenosine and to understand the link between cardiac energetics and adenosine formation. In 1974 there was a first International Symposium on Adenosine and Adenine Nucleotides which since then took place every four years. The proceedings of these meetings were published as books and give a good account on the progress made in the physiology, pharmacology, biochemistry, and clinical use of adenosine.[6,13,17,41,114] Cardiac adenosine metabolism has been reviewed by Schrader,[104] Sparks and Bardenheuer,[116] Belardinelli et al[12] and Olsson and Pearson.[92] There have been recent specialized reviews on the role of adenosine in pain during angina pectoris,[118] circadian variations of adenosine,[21] adenosine receptor subtypes,[68,91] the ionic basis of the cardiac action of adenosine,[14,113] myocardial protection[66] and the cardiovascular pharmacology of purines.[100]

In biochemical terms, adenosine is the dephosphorylated breakdown product of adenine nucleotides, and ATP is well known to serve a key role in cellular energetics. Conventionally it was assumed that lack of metabolizable substrates such as oxygen causes a gradual decrease of ATP and subsequent formation of adenosine. The close link between cellular energetics and the formation of adenosine and the strong dependence of coronary blood flow on myocardial oxygen consumption has led to the hypothesis of metabolic coronary blood flow regulation.[16] This concept states that adenosine (or another vasodilator) acts in a feed back loop tending to bring the mismatch between ATP synthesis and demand back to equilibrium. In the past, several reports provided evidence for a reciprocal relationship between the energy state of the myocardium and the formation of coronary vasodilators such as adenosine.[20,50,51] The precise relationship between cardiac energetics and adenosine formation and release, however, has not been elucidated until recently, and will be discussed later in this chapter.

In addition to the above described classical cardiac actions, adenosine was shown to reduce the hemodynamic and metabolic effects of catecholamines.[37,106] This antiadrenergic action of adenosine my prevent sympathetic overstimulation of the heart. More recently adenosine was demonstrated to attenuate postischemic dysfunction by delaying calcium overload[66] and to mediate ischemic preconditioning which renders the myocardium less vulnerable during a period of subsequent myocardial ischemia.[38] The extremely diverse actions of adenosine in the heart can functionally be grouped into those which increase oxygen delivery via enhanced coronary flow and those which reduce myocardial oxygen consumption. Both effects tend to improve the free energy of ATP hydrolysis and because of this overall regulatory effect, adenosine has been termed to be a retaliatory or homeostatic molecule.[88,105]

The findings that adenosine can be formed by almost all cells and organs studied and that adenosine receptors are ubiquitously distributed do not necessarily imply an important physiological role of this nucleoside. Whether adenosine is functionally relevant therefore asks an important quantitative question: Is the concentration of adenosine at the receptor site sufficient to cause receptor activation? Furthermore, what are the metabolic mechanisms that link adenosine to cardiac energy metabolism? One way of tackling this question is by using specific and potent adenosine receptor antagonists now available.[68,91] The other approach is to measure adenosine. Ideally this would be the concentration of adenosine in the interstitial space. However, there is no direct way to assess the interstitial concentration of adenosine and one has to rely on indirect measurements. The reasons for this difficulty are a direct consequence of the following findings:

1. The various cell types within the heart (cardiomyocytes, smooth muscle, endothelium, pericytes) show a different ability to form and degrade adenosine. The endothelium, e.g., constitutes a metabolic barrier for infused adenosine[85] and the degradation of adenosine down to inosine, hypoxanthine and uric acid occurs predominantly in this cell compartment.[11,26]

2. Aside from cellular compartmentation of adenosine metabolism an intracellular and extracellular site of production involving a cytosolic- and an ecto-5'-nucleotidase must be differentiated.[112] Intracellularly-produced adenosine reaches the extracellular space by a transporter which has been recently cloned.[47]

3. More then 95% of the adenosine content which is measured in the heart after conventional acid extraction is protein bound, most likely to SAH-hydrolase.[34,93] Because of this high background signal, measurement of changes of the free and biological active adenosine requires a different approach.[34]
4. The plasma half life of adenosine is extremely short and was estimated to be 0.6 sec in human blood plasma under control conditions.[80] Even when the plasma levels were increased to 1 μmol L^{-1} adenosine, the half life was still as short as 1.2 sec.
5. The adenosine metabolism is known to be complex. At a given steady state metabolic condition the intracellular concentration of adenosine is a function of the differences in the rates of adenosine production by two enzymatic reactions (5'-nucleotidase, SAH-hydrolase) and adenosine removal by another two enzymes (adenosine deaminase, adenosine kinase) and adenosine transport across the membrane. It should be noted that a metabolic cycle exists between AMP and adenosine which consumes energy.[60]

All these factors make the understanding of the metabolism of adenosine a rather complicated matter. As will be discussed below, from a regulatory point of view the various enzymes involved can be differently influenced making it a highly interesting control system by which the generation of a biologically active signal, namely adenosine, permits fine tuning of cell and organ function in order to keep cellular energetics in equilibrium.

Additional lessons learned over the past decade are that the dynamics of adenosine and adenine nucleotide metabolism can only incompletely be understood by chemically measuring tissue content of high energy phosphates and their breakdown products. This is because several of the important metabolites involved, such as ADP and AMP, are protein bound and thus do not reflect the biologically active concentration of these nucleotides. Furthermore changes in the concentration of a metabolite do not permit one to extrapolate to a flux rate through a given pathway, since aside from the formation rate (input) also the rate of removal/bioconversion (output) importantly influences the metabolite concentration. Equally well, measurements of in vitro enzyme activities do not permit to extrapolate to the in vivo situation, since many of the regulatory factors influencing enzyme activity in the intact heart are only incompletely understood and the substrate concentration is usually unknown.

Progress has been made in recent years by refining the methods for measuring the biologically free metabolite concentration of ADP, AMP and adenosine. In the case of ADP the free nucleotide concentration can be calculated assuming creatine kinase at equilibrium and measuring ATP, phosphocreatine (PCr) and creatine (Cr) chemically or more conveniently by ^{31}P NMR. NMR spectroscopy in addition permits the simultaneous measurement of intracellular pH where:

$$[ADP] = ([ATP] \cdot [Cr]) / ([PCr] \cdot K_{obs})$$

K_{obs} is the observed, pH-dependent equilibrium constant of the creatine kinase reaction.[67]

Similarly AMP can be calculated assuming myokinase at equilibrium where:

$$[AMP] = K_{MK} \cdot [ADP]2 / [ATP]$$

K_{MK} denotes the equilibrium constant of the myokinase reaction.[67]

Measurement of the free adenosine concentration makes use of the fact that in the presence of a saturating concentration of L-homocysteine, the rate of SAH accumulation is proportional to the free cytosolic adenosine concentration.[34] This technique has proven useful in measuring changes in the steady state concentration of adenosine.[29,63,72]

Another important concept is that between adenosine formation and adenosine release must clearly be differentiated. While adenosine formation denotes the rate of conversion from AMP and SAH to adenosine within the heart, the term adenosine release signifies the net release of adenosine with the coronary effluent perfusate or transudate which is the difference between its continuous production and metabolism. Further progress in the understanding of the dynamics of adenosine metabolism comes from the development of a comprehensive model of transport and metabolism of adenosine which incorporates our recent knowledge on cellular adenosine compartmentalization as well as all relevant enzyme activity and transport data.[61] Adenosine production rates in the heart can now be described in a quantitative manner and the adenosine model appears to have a predictive value.[32]

The scope of this review is to outline the key features of the metabolism of adenosine in endothelial cells and cardiomyocytes which are essential for understanding the various regulatory functions of adenosine in the heart. We will summarize recent findings demonstrating that the futile cycle between AMP and adenosine serves the important purpose to potentiate adenosine release in view of only small changes in free AMP.[29] Furthermore, evidence will be summarized showing that the rate of intracellular adenosine removal exceeds the intracellular production of adenosine, so that in the normoxic heart cellular uptake of adenosine dominates over release.[32]

OVERVIEW OF ADENOSINE METABOLISM AND ENERGY BALANCE

Because anaerobic energy production contributes only 10-13% to the normal ATP requirement[56,73] cardiac energy production strongly depends on oxygen, the substrate that may become limiting. As outlined in Figure 2.1, the ATP formed by oxidative phosphorylation in mitochondria is transported to the site of ATP consumption, the contractile proteins actively performing work. As a consequence of ATP hydrolysis, ADP is formed which in the presence of oxygen becomes rephosphorylated. Cytosolic AMP is formed from ADP by the activity of the myokinase reaction which is assumed to be at equilibrium under most conditions. Between AMP and adenosine there is an important substrate cycle which is catalyzed by cytosolic-5'-nucleotidase and adenosine kinase. Details of this substrate cycle will be discussed later. Adenosine can also be formed by the transmethylation pathway involving

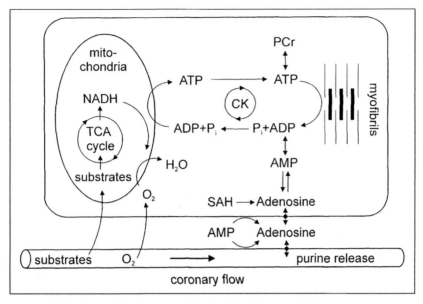

Fig. 2.1. Overview of cardiac energy and adenosine metabolism.

the hydrolytic cleavage of S-adenosylhomocysteine (SAH). Figure 2.1 also illustrates that adenosine can be formed extracellularly by a cascade of ecto-nucleotidases which have been identified on myocardial and vascular cells. Adenosine has a high membrane permeability and is actively degraded to inosine and hypoxanthine especially by endothelial cells.[11] When produced at an accelerated rate, adenosine is rapidly removed either by metabolic inactivation and/or washout from the heart by the coronary circulation. As will be discussed later, this makes the cardiac adenylate pool an open system which may serve to improve the free energy of ATP hydrolysis.[62]

In principle there are three sources of adenosine in the heart: cardiomyocytes, endothelial cells and vascular smooth muscle cells. From a quantitative point of view cardiomyocytes comprise the largest cellular compartment while vascular cells account only for about 3-5% of the heart cell volume. During cardiac hypoxia with substantial ATP breakdown, most of the adenosine is derived from cardiomyocytes. Under normal physiological conditions, however, endothelial cells have been calculated to contribute up to 13-25% to coronary venous adenosine release.[42,64] More recent evidence suggests that the concentration of adenosine in the coronary venous effluent perfusate of the guinea pig heart could be explained solely by endothelial adenosine metabolism.[75]

There are also major differences in the metabolic capacity of cardiac cells to degrade adenosine. Available evidence suggests that not only in rodent but also human heart adenosine deaminase and purine nucleoside phosphorylase are predominantly localized in endothelial cells[11,57] while the presence of xanthine oxidase is species dependent and low in the human heart.[26] There are also substantial differences in the in vitro activity (V_{max}) of key enzymes of adenosine metabolism between various species. For example, the human heart usually displays lower activities of 5'-nucleotidase, adenosine deaminase and nucleoside phosphorylase while adenosine kinase is similar to that of rodent heart.[57,119] Although these species differences are interesting, no conclusion can be drawn from these data as to the rate of adenosine formation in vivo. This would require knowledge of the substrate concentration and determination of flux rates in the intact heart in which the respective enzymes most likely operate well below K_M rather than close to V_{max}.

PATHWAYS OF ADENOSINE PRODUCTION
IN NORMOXIA AND HYPOXIA

The major source of adenosine even in the normoxic heart is dephosphorylation of AMP by action of *cytosolic 5'-nucleotidase*. Recent studies in the isolated heart demonstrate that the flux through the AMP → adenosine pathway by far exceeds the other adenosine producing pathways.[29,60] Adenosine formation by the hypoxic heart occurs intracellularly since inhibition of ecto-5'-nucleotidase with a specific antibody[77] or pharmacologic inhibition of the enzyme with AOPCP[112] were ineffective to reduce adenosine release. Similarly, nucleoside transport inhibitors attenuated hypoxic adenosine release. An AMP-specific cytosolic 5'-nucleotidase has been purified in the dog heart which is allosterically regulated: ADP and Mg^{2+} serve as activators while the enzyme is inhibited by

ATP.[24] It is conceivable that in the course of tissue hypoxia characterized by ATP breakdown, not only is the substrate concentration of AMP increased but 5'-nucleotidase becomes deinhibited by ADP and Mg^{2+}. As will be discussed later it appears that these two effects can not quantitatively explain the hypoxia-induced increase in the release of adenosine.

Adenosine can be formed extracellularly by action of *ecto 5'-nucleotidase* which is particularly active on endothelial cell surface.[84] In the heart, however, there are substantial species differences[90] which make a functional interpretation of this enzyme difficult. It is also unclear at present from which cellular source extracellular AMP is derived. Platelets and endothelial cells are known to release adenine nucleotides.[18] However, to what extent intact cardiomyocytes can release ATP, ADP or AMP and under which circumstances has not been defined.

S-Adenosylhomocysteine (SAH) hydrolase is a cytosolic enzyme which catalyzes the reversible hydrolysis of SAH to adenosine and L-homocysteine.[109] This production of adenosine from SAH is directly proportional to the rate of transmethylation reactions which make use of S-adenosylmethionine (SAM) as methyl donor. In the rabbit, coronary microvessels displayed a four-fold higher activity compared with isolated cardiomyocytes,[79] while in the guinea pig and dog heart global SAH-hydrolase was similar to that of isolated cardiomyocytes.[19] This enzyme is also present in the human heart.[19] The flux through the transmethylation pathway is not altered by cardiac hypoxia.[35,71]

Adenosine deaminase catalyzes the conversion of adenosine to biologically inactive inosine. This enzyme is predominantly localized in endothelial cells.[11,83] Its high K_M value (70 µmol L^{-1}) suggests that the flux through this pathway only becomes important when intracellularly adenosine increases above the K_M-value of adenosine kinase which is around 1 µmol L^{-1}.

Adenosine kinase is a low K_M intracellular enzyme which effectively salvages any adenosine formed intracellularly and adenosine transported into the cytosol following its extracellular formation. More than 95% of the adenosine formed from AMP is converted back to AMP thereby creating an energy consuming substrate cycle. Adenosine kinase was recently reported to be inhibited by P_i[45] and the hypoxia-induced rise in P_i may in part be responsible for the accelerated adenosine release due to hypoxia-induced inhibition of adenosine kinase.[75] As will be discussed later intracellular rephosphorylation by adenosine kinase forms a sink for adenosine so that the concentration gradient of this nucleoside in the well-perfused heart is from extra- to intracellular.[32] The high activity of adenosine kinase in all cardiac cells is also responsible for the finding that endothelial adenine-nucleotides can be selectively prelabeled within the intact heart by intracoronary infusion of radioactive adenosine.[83,107] The technique of prelabeling of the endothelial cell compartment is an elegant way to measure in otherwise intact hearts the contribution of endothelial adenosine release relative to that of the total heart.[8,11,36,64]

THE ENDOTHELIUM AS A METABOLIC BARRIER FOR ADENOSINE

Endothelial cells lining the vascular space have important roles in the transport of adenosine between the vascular and the interstitial region and for the interactions between blood components and the vascular wall. In capillaries, the endothelial cells represent the only barrier between the vascular and the interstitial space (Fig. 2.2). It is generally assumed that adenosine can cross between these two regions either via diffusion through the interendothelial clefts or via facilitated diffusion across the luminal and abluminal endothelial cell membrane with interposed molecular diffusion through the cytoplasm. The resistance that impedes diffusion of adenosine across the endothelial cells is predominantly due to a metabolic barrier. Thus, transport and diffusion processes act in concert with the metabolic features of the endothelial cells. In the following will be reviewed our present understanding of these processes. In doing so we will follow the evolution from the most reductionist to integrated models.

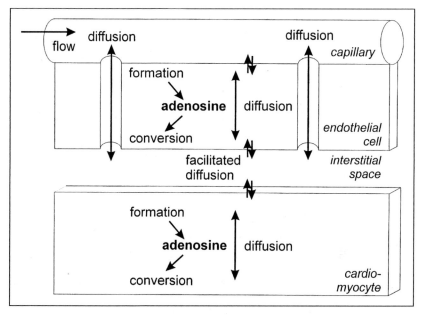

Fig. 2.2. Overview of processes involved in adenosine transport and metabolism in the microcirculation. In addition to the processes indicated there is also axial diffusion (along the capillary length) which has not been included in this scheme.

IN VITRO ENZYME ANALYSIS

The earliest quantitative information suggesting the possible importance of the vascular endothelium for whole tissue adenosine metabolism comes from measurements made by Nees et al in 1980.[86] These authors studied the concentration of purine metabolites and the activity of several enzymes involved in cardiac adenine metabolism. One of the most striking results was that in coronary endothelial cells concentrations of adenine nucleotides as well as those of adenosine, inosine, hypoxanthine and adenine exceeded those of whole myocardial tissue between three-fold (adenine nucleotides) and 100-fold (purines). Coronary endothelial cells were found to exhibit particularly high activities of ecto-5'-nucleotidase, purine nucleoside phosphorylase, adenosine kinase and adenosine deaminase.[30,83,84] The functional relevance of these results for a better understanding of global cardiac adenosine metabolism is not fully clear.

Adenosine Metabolism and Transport
in Isolated Endothelial Cells

Isolated endothelial cells exhibit two functional metabolic cascades that are important. One is related to the production, the other to the metabolism of adenosine. Ecto-nucleotidases present on endothelial cells catalyze the rapid conversion of ATP → ADP → AMP → adenosine.[97] In pig aortic endothelial cells the rate limiting step in this metabolic sequence seems to be the dephosphorylation of 5'-AMP to adenosine catalyzed by ecto-5'-nucleotidase.[97] As the K_M-values of the ecto-nucleotidases of endothelial cells are between 50 µmol L^{-1} (AMP) and 350 µmol L^{-1} (ATP) and the extracellular adenine nucleotide concentrations are physiologically in the nanomolar range, increases of the extracellular adenine nucleotide concentration are expected to result in proportional increases of extracellular adenosine production. Endothelial cells appear to release adenine nucleotides continuously. Pearson et al prelabeled the adenine nucleotide pool of pig aortic endothelial cells with ³H-adenosine and showed that after a washout period the label was continuously released with the adenine nucleotide fraction.[98] The steady-state release of labeled adenine nucleotides was enhanced when exposing cells to trypsin or thrombin. Deussen et al[33] provided evidence that pig aortic endothelial cells superfused in a perfusion column liberate native adenine nucleotides at a rate quantitatively similar to that of adenosine (Fig. 2.3). In this model a considerable fraction of the adenine nucleotides released by the cells was converted to adenosine within seconds during the column passage. Inhibition of the ecto-5'-nucleotidase with α,β-methylene adenosinediphosphate (AOPCP) enhanced the recovery of adenine nucleotides and decreased the concentration of adenosine in the column effluent perfusate. More recent experiments indicate that continuous extracellular adenosine production from 5'-AMP via ecto-5'-nucleotidase occurs also in coronary endothelial cells from guinea pig heart.[18,75] Although there is good evidence for continuous release of adenine nucleotides from the endothelial cell, the mode by which the adenine nucleotide(s) is (are) released remains a matter of debate. The multidrug resistance (mdr1)

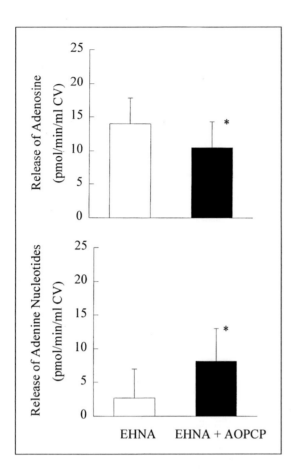

Fig. 2.3. Release of adenosine and adenine nucleotides from aortic endothelial cells. EHNA = erythro-9-(2-hydroxy-3-nonyl)adenine, inhibitor of adenosine deaminase; AOPCP = α,β-methylene adenosinediphosphate, inhibitor of ecto-5'-nucleotidase. *P<0.05 vs. EHNA, data are from n = 11 experiments. Modified with permission from Deussen A et al, Am J Physiol 1993; 264: H692-H700.

gene product, a P-glycoprotein, which has been suggested to mediate cell nucleotide efflux[1] has not been documented for cardiac tissues.[23,120] Thus, the mode of adenine nucleotide release from endothelial cells awaits clarification in future studies.

Extracellular adenosine is effectively taken up by the endothelial cell.[97] In pig aortic endothelial cells two distinct transport sites have been described with K_M-values of 3 and 250 µmol L^{-1}, respectively, for adenosine. The low affinity transporter has a V_{max} that exceeds that of the high affinity transporter approximately three-fold.[97] Thus, under physiological conditions (adenosine concentration in the nanomolar range) the high affinity transport site, due to its low K_M, is responsible for more than 95% of cellular adenosine uptake. Coronary endothelial cells from guinea pig heart are preferentially rich in ^{3}H-nitrobenzylthioinosine binding sites[95] which suggests that the endothelial cell membrane transporter may be classified as a facilitative, nonconcentrating type.[96] Using an endothelial cell column perfusion system[33] it was found that blockade of adenosine membrane transport with nitrobenzylthioinosine doubles adenosine release from the cells.[76] This suggests that in this experimental model approximately 50% of the adenosine produced outside the coronary endothelial cell under control conditions is taken up into the cytosol. The corollary is that the transmembraneous adenosine concentration gradient is from extra- to intracellular.

The second metabolic cascade in the endothelial cell serves the metabolism of adenosine to uric acid (adenosine → inosine → hypoxanthine → xanthine → uric acid).[11,42] While metabolism of adenosine to hypoxanthine is found in coronary endothelium of all mammalian species, the metabolism of hypoxanthine to xanthine and uric acid is largely species dependent.[26,82] Uric acid is the dominant catabolite of adenosine in species which exhibit a high xanthine oxidase activity, e.g., mouse, rat and guinea pig. In human and rabbit heart uric acid is produced from hypoxanthine at a considerably lower rate. In pig heart there is almost no uric acid detectable in the coronary venous effluent perfusate even during continuous arterial infusion of hypoxanthine.[26] From comparative experiments carried out on coronary endothelial cells from guinea pig heart and macrovessels (aorta, pulmonary artery, caval vein) from pig, it has been concluded that endothelial cells which originate from different vascular segments differ in their ability to take up and metabolize adenosine.[42] However, as the endothelial cells were also obtained from different animal species, it cannot be readily excluded that observed differences were rather due to a difference by species than to a difference by vascular origin.

In addition to these two metabolic cascades which are connected via transmembraneous exchange of adenosine there is intracellular formation of adenosine from AMP and S-adenosylhomocysteine (see above). Besides deamination of adenosine, another route of metabolism is rephosphorylation of adenosine to AMP. Quantitative analyses of enzyme inhibitor experiments carried out on pig aortic endothelial cells indicate that 70-90% of the global cellular adenosine production rate must be attributed to intracellular production sites.[33,76] However, due to the low K_M and a sufficiently high V_{max} of the adenosine kinase, intracellular phosphorylation of adenosine is so effective that not only the adenosine produced within the cytosol, but also a fraction of the adenosine produced in the extracellular region are phosphorylated to AMP. This causes the normal adenosine concentration gradient across the endothelial cell membrane to fall from the extracellular to the intracellular space, the motor of this gradient being the high activity of endothelial cell cytosolic adenosine kinase. The implications of this effective competition of adenosine kinase for cytosolic adenosine are described in the following.

METABOLIC AND TRANSPORT FEATURES OF ENDOTHELIAL CELLS WITHIN THE HEART

Perfusion of the isolated guinea pig heart with ^{3}H-adenosine at submicromolar concentrations leads to deposition of the label superimposing the position of the endothelial cells.[83,85] Neither vascular smooth-muscle cells nor cardiomyocytes reveal uptake of tracer according to these autoradiographic studies. To understand the selective trapping of the label in endothelial cells in a quantitative manner it is pertinent to review the results obtained from indicator dilution experiments. Such experiments which use bolus injections of labeled compounds are particularly helpful in quantifying the metabolism of endothelial cells within an intact circulation. Multiple indicator dilution experiments are based on the use of multiple control tracers which are coinjected but differ with respect to their volumes of distribution, their rates of permeation through capillary wall or cell membranes and their rates of metabolism.[10] Appropriate tracer controls are those giving refined information of a specific part of the coronary system that is relevant to transport and metabolism of the test substance adenosine. As the endothelial cell region is located adjacent to the capillary region, analysis of fractionally collected label from the venous effluent perfusate offers a high sensitivity toward endothelial processes.[9] In a typical experiment ^{3}H-adenosine is coinjected with ^{131}I-albumin which serves as a marker of the capillary region and ^{14}C-sucrose which passes through the interendothelial clefts and therefore distributes over the entire extracellular space.[124] Displacement of the coronary venous outflow characteristics of the ^{14}C-sucrose from that of ^{131}I-albumin is then interpreted as resulting from transcapillary exchange, while differences between the ^{3}H-adenosine and ^{14}C-sucrose are indicative of membrane transport and metabolism of adenosine (Fig. 2.4). If metabolism of adenosine to inosine is not inhibited, the venous outflow profiles of reaction products resulting from the metabolic cascade down to uric acid have to be interpreted individually.[59] Indicator dilution experiments carried out in the isolated perfused guinea pig heart show that intravascular adenosine is twice as likely to enter the endothelial cells than to permeate the clefts.[124] The underlying mathematical model analysis further indicated that once tracer had entered the cell it was more likely to undergo reaction than to escape from the cell. This was evident from the fact that the endothelial cell intracellular clearance rate of adenosine was calculated to be several-fold higher than the rate of permeation of adenosine across the luminal endothelial cell membrane. Given the velocity of the kinetics during a single capillary passage and the fact that homocysteine was absent from the perfusion buffer (SAH-hydrolase reaction strongly favoring hydrolysis of S-adenosylhomocysteine) the observed disappearance of adenosine could only have taken place via adenosine kinase or adenosine deaminase. Information as to whether the adenosine kinase or adenosine deaminase pathway was more important in metabolizing the adenosine fraction in in situ endothelial cells can be obtained by summing the different fractions of labeled metabolites which appear in the coronary effluent perfusate. From the analysis of indicator dilution experiments such as shown in Figure 2.4 it was found that 62% of the tracer adenosine was retained in the heart, while 20% of the labeled adenosine emerged unchanged; only 4% was recov-

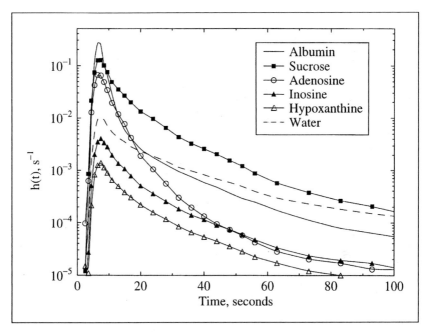

Fig. 2.4. Coronary venous outflow curves of adenosine and metabolic products after bolus injection of adenosine into the coronary inflow of an isolated perfused guinea pig heart. The adenosine was coinjected with albumin (vascular reference) and sucrose (extracellular reference). Reprinted with permission from Kroll K et al. In: Bassingthwaighte JB et al, eds. Whole Organ Approaches to Cellular Metabolism. New York: Springer, 1998.

ered in the venous outflow as labeled inosine, 1% as hypoxanthine plus xanthine and 13% as tritiated water, the final product of the metabolism of tritiated adenosine.[58]

The dynamic of the vascular endothelium for the clearance of vascular adenosine has recently been shown most strikingly by indicator dilution experiments performed in the dog heart in situ.[65] In Figure 2.5 results from coinjections of albumin, sucrose and adenosine are shown. In contrast to the experiments conducted on the isolated perfused guinea pig heart (Fig. 2.4) it is obvious that the adenosine outflow curve is totally flat; only with a 100-fold magnification the adenosine tracer transient is visible. This indicates that first-pass extraction of adenosine in the canine coronary circulation is almost complete and that the local uptake of

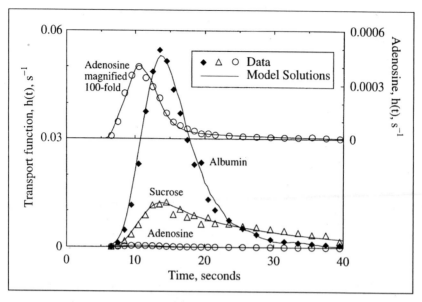

Fig. 2.5. Multiple indicator dilution curves for tracer adenosine, sucrose and albumin coinjected into the coronary artery of a dog. Blood samples were taken from the coronary sinus and analyzed for tracer concentration. The main panel shows adenosine, sucrose and albumin curves using the scale on the left hand ordinate. The adenosine curve is also shown at 100-fold magnification using the scale on the right hand ordinate. Reprinted with permission from Stepp DW et al, Circ Res 1996; 79:601-610.

adenosine is flow limited. This flow limitation of adenosine extraction is also evident from a second striking feature of this adenosine dilution curve, namely the fact that the peak of the adenosine dilution curve preceeds the peak of the intravascular volume marker albumin. The explanation for the shorter transit time of adenosine is spatial blood flow heterogeneity which is a physiological feature of left ventricular myocardial perfusion.[10,115] Only in myocardial regions with a local blood flow several-fold above the average flow, the coronary transit time is obviously too short to permit complete tracer adenosine extraction. Since those high flow regions, however, contribute only a small mass to the entire left ventricular myocardium, the albumin fraction passing through these regions is negligible and does not significantly contribute to the venous outflow curve. In the in vivo dog heart 85%

of the injected adenosine was retained in the heart, while 14% was recovered as membrane permeable metabolites.[65] The reasons for these quantitative differences between the in situ dog heart and the isolated perfused guinea pig heart (see above) are as follows:

1. Coronary flow rate in the blood-perfused dog heart is 10-fold lower than that in buffer-perfused guinea pig heart. Therefore the ratios between the permeability-surface area product of the endothelial cell luminal membrane and the endothelial cell clearance rate, respectively, and myocardial blood flow are high which favors cellular sequestration of the tracer.

2. The endothelial barrier is more leaky in isolated preparations which is mainly due to the absence of physiological concentrations of plasma proteins. This results in moderate interstitial edema in isolated hearts associated with a widening of interendothelial clefts. As a result, the likelihood for adenosine to permeate via the cleft and to redistribute back into the vascular region is augmented as compared with the likelihood to be transported into the endothelial cell via the luminal membrane and to undergo immediate biochemical reaction.

In conjunction with early prelabeling studies[107] these indicator dilution experiments indicate that the adenosine kinase pathway is normally the major route for adenosine metabolism in in situ endothelial cells. This is emphasized by the mathematical modeling results shown in Figure 2.6. Analysis of the adenosine outflow curve after bolus injection into the canine coronary artery reveals an endothelial clearance rate of 82 ml min^{-1} g^{-1}. Treatment with an adenosine deaminase blocker (EHNA) reduces this clearance rate to 55 ml min^{-1} g^{-1}, while treatment with an adenosine kinase blocker reduces it to 12 ml min^{-1} g^{-1}. During combined blockage of adenosine kinase and adenosine deaminase, the adenosine outflow curve is almost identical to that of sucrose and the analysis reveals an endothelial clearance rate of only 0.6 ml min^{-1} g^{-1}. These analyses show that in in situ canine coronary endothelial cells the cytosolic adenosine is approximately 5-fold more likely to be phosphorylated by the adenosine kinase reaction than

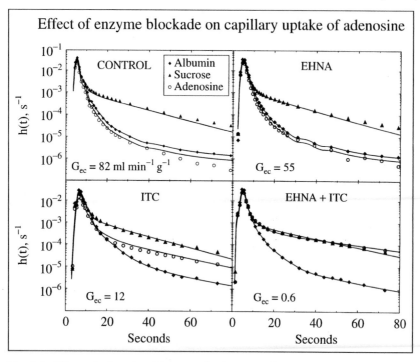

Fig. 2.6. Effect of adenosine kinase and adenosine deaminase inhibition on the endothelial clearance rate of adenosine (G_{ec}). The clearance rate was calculated from fitting the coronary venous outflow curve of adenosine in the dog heart after arterial bolus injection. EHNA = erythro-9-(2-hydroxy-3-nonyl)adenine, inhibitor of adenosine deaminase; ITC = iodotubercidine, inhibitor of adenosine kinase. Reprinted with permission from Kroll K et al. In: Bassingthwaighte JB et al, eds. Whole Organ Approaches to Cellular Metabolism. New York: Springer, 1998.

to undergo deamination to inosine. The high activity of adenosine kinase provides an explanation for the rapid kinetics of adenosine clearance in the microcirculation[65] and the selective autoradiographic labeling of the endothelial cells after intracoronary infusion of labeled adenosine.[85] It should be noted, however, that the situation in humans may be more complex than indicated by the results reviewed here. In contrast to canine blood, human erythrocytes have a high capacity membrane transport system for adenosine[22] and the in vitro half-life of adenosine in human blood plasma is approximately 1/360 of that in canine plasma.[80] Thus, it is to be expected that human erythocytes significantly compete with endothelial cells for capillary adenosine.

Contribution of Endothelial Cells
to Global Cardiac Adenosine Production

The selective uptake of ^{3}H-adenosine into coronary endothelial cells has been used experimentally to label the endothelial cell adenine nucleotide pool and, by measurement of the specific radioactivity of the released adenosine, to obtain insight into the contribution of in situ endothelial cells to global cardiac adenosine release.[8,36,64] The fact that in those experiments following completion of the prelabeling procedure a steady-state release of radioactive adenosine is observed, can be taken as evidence that coronary endothelial cells contribute to the global cardiac adenosine release under well-oxygenated conditions. As, however, the specific radioactivity of the adenosine released with the coronary effluent perfusate declines during stimulation of cardiac adenosine production, e.g., during hypoxic perfusion,[36] hypercapnic perfusion,[36] perfusate acidosis,[8] infusion of catecholamines[8] or infusion of acetylcholine[8,36] this was taken to indicate that the relative contribution of the endothelium to global adenosine release is decreased under conditions of enhanced production. Using information on the specific radioactivity of the endothelial cell cAMP pool, Kroll et al[64] calculated the contribution of the endothelium to coronary venous release to be 14% in the guinea pig heart. An estimate of 25% has been reported by Becker and Gerlach[11] using a different experimental approach. It should be emphasized that for methodological reasons these data do not reveal the contribution of the endothelial cell to global cardiac adenosine production as opposed to global release. The contribution of the endothelial cells to the global cardiac production rate may actually be smaller, as the contribution of the endothelial cell region to coronary venous release may be amplified due to the direct vicinity of the endothelial cells to the vascular region. On the other hand, recent comparative studies conducted on coronary endothelial cells from guinea pig heart and isolated perfused hearts indicate that endothelial adenosine metabolism can quantitatively account for the adenosine concentration of the coronary venous effluent perfusate.[75]

IMPLICATIONS OF THE ENDOTHELIAL METABOLIC BARRIER

The dose-response curve for exogenous adenosine has been determined from intracoronary infusion of adenosine and measurement of the coronary flow response in the isolated perfused guinea pig heart.[108] Threshold coronary vasodilation was at an intracoronary concentration of 10^{-8} mol L^{-1} adenosine and a maximal flow effect resulted at adenosine concentrations above 10^{-6} mol L^{-1}. That is, the full dose-response curve extended over two orders of magnitude. Since it is now well established that the endothelium acts as a metabolic barrier for adenosine, this apparent dose-effect relationship needs reinterpretation for two reasons. (1) As described above, the high metabolic activity of the endothelium significantly influences the vascular as well as the interstitial concentration of adenosine. (2) Adenosine receptors have been described on vascular smooth muscle cells and endothelial cells.[31,44,81] Given the recent findings that stimulation of endothelial cells may result in the liberation of vasoactive compounds which in turn act on the vascular smooth-muscle cell (NO, prostaglandins, endothelin, etc.) it remains to be clarified whether adenosine acts via endothelial or smooth-muscle receptors to induce the decrease of coronary vascular resistance.

First information on the true dose-response curve for adenosine in the canine coronary circulation has been obtained in experiments conducted on the dog heart in situ.[117] Adenosine was either applied via the coronary arteries or intracellular adenosine metabolism via adenosine kinase and adenosine deaminase was blocked. From the measured coronary venous adenosine release the concentration of adenosine was calculated for the capillary region and the interstitial region using a mathematical model that had previously been calibrated against results obtained in indicator dilution experiments using the same experimental preparation.[65] It was found that in the coronary circulation of the dog heart the dose response curve was surprisingly steep with a threshold arterial concentration of 0.7 µmol L^{-1} and a half maximal effect at 136 µmol L^{-1}. Moreover, calculating the adenosine concentration in the interstitial region for the two experimental conditions (arterial infusion vs. enzyme block) it was found that both dose-response curves were superimposable. A similar analysis was conducted on our own data obtained on the coronary venous adenosine release rate in the isolated perfused guinea pig heart under conditions of hypoxic perfusion and exogenous adenosine infusion (Fig. 2.7). Regional adenosine concentrations were calculated by a mathematical model[61] calibrated against experimental data obtained in the same preparation.[34] In the upper panel of the figure the adenosine concentrations are shown for the capillary, the interstitial and the parenchymal cell region for the condition of hypoxic perfusion (gradual reduction from 95-15% O_2 in the equilibrating gas). In the lower panel are shown those concentrations for intracoronary infusion of adenosine. As is to be expected, with hypoxic perfusion when the venous outflow concentration is fitted by adjusting the cellular adenosine production rates the calculated concentration gradient is decreasing from the parenchymal cell to the interstitial to the capillary region (from right to left in Fig. 2.7). However, as shown in the lower panel for infusion of exogenous adenosine, simulated by enhancing the model input concentration in the arterial vessel segment, the concentration was falling from the capillary to the interstitial to the parenchymal cell region. That is, the order of the dose-response curves for the different tissue regions is reversed. (Only for very low inflow adenosine concentrations the mean capillary concentration was below that of the interstitial region.) The important result concerning the receptor site that may be responsible for the observed vasodilation is the finding that only the position of the dose-response curve calculated for the interstitial region was independent of the intervention used to increase the adenosine concentration. Most notably, for the capillary region the dose-response curves calculated for the two interventions are clearly dissociated suggesting that an adenosine receptor located on the luminal endothelial membrane is most likely not responsible for the observed vasodilation.

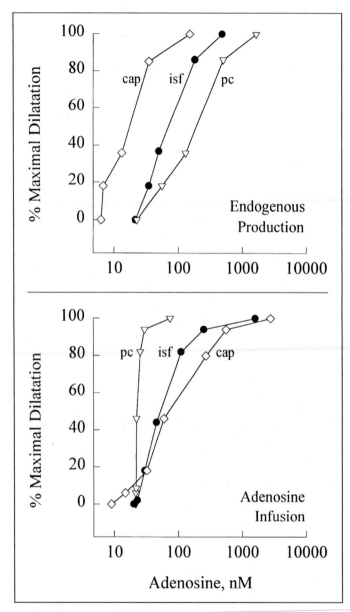

Fig. 2.7. Dose-response curves of adenosine calculated from fitting the coronary venous release rates of adenosine by a mathematical model of adenosine metabolism. The dose-response curves were calculated for the capillary, the interstitial and the parenchymal cell region. The curves calculated for the interstitial region are superimposable for the conditions of intracoronary adenosine infusion and endogenous production during graded reduction of PO_2 in the perfusion buffer. For further information see text.

CARDIAC ENERGY STATUS
AND ADENOSINE METABOLISM

In the well-oxygenated heart, the high turnover of the cytosolic ATP pool requires a near-perfect match between ATP formation and consumption, maintaining constant cytosolic concentrations of both ATP (5-10 mmol L^{-1}), ADP (40-80 µmol L^{-1}) and inorganic phosphate (1-2 mmol L^{-1}). Due to the equilibrium of the creatine kinase reaction (PCr + ADP + H^+ ⇔ Cr + ATP), the ATP/ADP ratio is reflected in the phosphocreatine / creatine ratio (10-15 and 5-10 mmol L^{-1}, respectively). Cytosolic free AMP (100-400 nmol L^{-1}), the immediate precursor of adenosine, is formed from ADP in the myokinase reaction (2 ADP ⇔ AMP + ATP) which is generally assumed to be in equilibrium. Thus under physiological conditions the rates of ATP synthesis and degradation are virtually identical and a high level of the free energy of ATP hydrolysis (ΔG_{ATP}) is maintained (-60 to -65 kJ mol^{-1}). The latter is the chemical potential energy released by the hydrolysis of ATP that is available to the cell for performing work.

Whenever ATP degradation exceeds ATP formation, free cytosolic ADP and in consequence free AMP increase, giving rise to an enhanced rate of adenosine formation, to an elevation of cytosolic and interstitial adenosine concentrations and ultimately to a rise in adenosine release. Since the myocardium can adapt to an enforced reduction in ATP formation by downregulating ATP breakdown ("myocardial hibernation") (e.g., refs. 4,54,101,110) this mismatch between ATP supply and demand is present for only a short period of time. Subsequently the heart may be operating at very similar concentrations of ATP but at a significantly lower level of ΔG_{ATP}. In line with this general concept, a rise in free cytosolic AMP in the absence of major changes in ATP has been reported to be associated with an increase in adenosine release in a variety of studies.[28,29,46,49,53] The changes in adenosine release were generally much greater than the reported changes in free AMP. While it has been known for a long time that adenosine is a very sensitive indicator of disturbances of the cardiac energy status, it is only recently that we have begun to understand the molecular and kinetic mechanisms responsible for the high sensitivity of the cardiac adenosine system.

In the following, we consider: (a) the importance of free cytosolic AMP in adenosine formation in the normoxic heart and compare it to the degradation of S-adenosyl-homocysteine (SAH) and extracellular AMP; (b) the mechanism translating small changes in free cytosolic AMP into major changes in cytosolic adenosine and ultimately adenosine release; and (c) the metabolic consequences of AMP degradation to adenosine with respect to the cardiac energy status when ATP turnover is impaired.

ROLE OF CYTOSOLIC AMP IN ADENOSINE FORMATION IN THE NORMOXIC HEART

As already described above (see Fig. 2.1), the cytosolic adenosine concentration is dependent on three input and three output pathways. Firstly, the dephosphorylation of free cytosolic AMP to adenosine is catalyzed by cytosolic 5'-nucleotidase. The second intracellular input to the adenosine pool is the transmethylation pathway (S-adenosyl-methionine → S-adenosyl-homocysteine L-homocysteine + adenosine), which due to the low intracellular homocysteine concentration physiologically results in net formation of adenosine.[27,35,70] Adenosine is also produced, albeit at a much smaller rate, from extracellular AMP by ecto-5'-nucleotidase[18,32] (see below) and may subsequently enter the cell by the well-described symmetric adenosine transporter.[99] Intracellular adenosine can subsequently be deaminated to inosine by adenosine deaminase, rephosphorylated to AMP by adenosine kinase or released from the cell. Due to the high activity of adenosine kinase and adenosine deaminase in the heart[57] it was assumed that a great proportion of adenosine formed will be immediately metabolized. In consequence, adenosine formation is likely to be much higher than coronary venous adenosine release, and venous release is not a good measure of adenosine formation.

To estimate total adenosine formation in the normoxic, saline perfused heart, the following approach has been widely em-

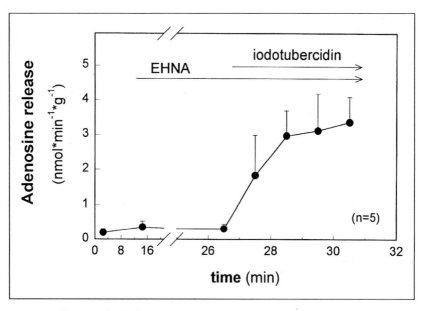

Fig. 2.8. Effect of blockade of adenosine deaminase (5 μmol L⁻¹ EHNA) and adenosine kinase (10 μmol L⁻¹ iodotubercidin) on coronary venous adenosine release in the well-oxygenated guinea pig heart. During blockade of both adenosine deaminase and adenosine kinase coronary adenosine release in the steady state is a measure of total cardiac adenosine formation (for details see text). Adapted with permission from Kroll K et al, Circ Res 1993; 73:846-856.

ployed:[33,40,60,89] Effective pharmacological blockade of adenosine metabolism by adenosine kinase (10 μmol L⁻¹ iodotubercidin) and adenosine deaminase (5 μmol L⁻¹ EHNA) will increase cytosolic adenosine. Since the maximal velocity of adenosine transport across the cell membrane is very high and not rate limiting in the steady state, all adenosine formed will be shunted towards release. In the normoxic heart the concentration of cytosolic adenosine is low (40-100 nmol L⁻¹),[34] and the rate of AMP dephosphorylation by either cytosolic or ecto 5'-nucleotidase as well as the SAH-hydrolysis rate are essentially independent of the adenosine concentration. Thus in the presence of effective blockade of adenosine deaminase and adenosine kinase the rate of adenosine release in the steady state is a valid measure of the rate of adenosine formation.

One example for this approach is depicted in Figure 2.8. Coronary venous adenosine release was 0.196 ± 0.045 nmol min⁻¹ g⁻¹

under basal conditions. Blockade of adenosine deaminase by EHNA increased adenosine release 2.3-fold. Additional blockade of adenosine kinase by iodotubercidin further raised adenosine release 10-fold to a value of 3.36 ± 0.71 nmol min^{-1} g^{-1}.[60] As deduced above, this release rate represents a good estimate of total cardiac adenosine formation. Similar data have been reported by Ely et al,[40] Deussen et al[32] and Decking et al.[29] Since in each study adenosine release rose up to 20-fold, formation of adenosine in fact by far exceeds release of adenosine. This suggests that the enzyme mainly responsible for global cardiac adenosine metabolism is adenosine kinase which is similar to the findings in endothelial cells (see discussion above).

In order to determine the relative contribution of cytosolic AMP, cytosolic SAH and extracellular AMP to global adenosine formation the following strategies were employed. Regarding extracellular adenosine formation, the release of adenine nucleotides in the normoxic guinea pig heart after blocking ecto 5'-nucleotidase by AOPCP was 0.43 ± 0.12 nmol min^{-1} g^{-1}.[18] This represents an upper estimate for extracellular AMP degradation, i.e., about 13% of the adenosine formation rate (3.36 nmol min^{-1} g^{-1}). When inhibiting adenosine transport by 1 µmol L^{-1} dipyridamole, adenosine release increased from 0.04 to 0.19 nmol min^{-1} g^{-1}.[32] This can be taken as a lower limit for extracellular adenosine formation. The fact that extracellular adenosine formation in the well-oxygenated guinea pig heart exceeds adenosine release has two important implications. (1) There is net inward flux via the symmetric adenosine transporter. (2) The interstitial adenosine concentration is higher than the cytosolic concentration in endothelial cells and cardiomyocytes.

Two separate approaches were used to estimate the rate of adenosine formation from SAH. Using radioactive tracer techniques, a SAH hydrolysis rate ranging from 0.62-1.35 nmol min^{-1} g^{-1} was calculated for guinea pig hearts.[70] However, when total adenosine formation was determined in this species (3.36 nmol min^{-1} g^{-1}) in absence and presence of SAH-hydrolase blockade, no difference in total adenosine formation was observed, suggesting adenosine formation from SAH to be well below 1 nmol min^{-1}

g^{-1}.[60] Measurement of the SAH accumulation rate in the presence of SAH hydrolase inhibition revealed an adenosine formation rate from SAH of 0.16 nmol min^{-1} g^{-1}.[35] This value can be taken as a lower limit for SAH hydrolysis in the normoxic guinea pig heart. Using the same experimental approach adenosine formation from SAH in dog myocardium was recently estimated to be 0.072 nmol min^{-1} g^{-1}.[72] Taken together, in the normoxic heart, about 10% of adenosine is formed extracellularly and about 10-25% is formed from SAH. The great majority of adenosine, accordingly, is formed from free cytosolic AMP via cytosolic 5'-nucleotidase.

AMP-ADENOSINE METABOLIC CYCLE IN THE NORMOXIC HEART

Since cytosolic AMP is both a precursor (via 5'-nucleotidase) and a product of adenosine (via adenosine kinase), these enzymes formally constitute a substrate cycle between AMP and adenosine. This was described by Arch and Newsholme as early as 1978.[5] Based on the kinetic constants (V_{max} and K_M) of 5'-nucleotidase (2420 nmol min^{-1} g^{-1}; 137 μmol L^{-1}) and adenosine kinase (136 nmol min^{-1} g^{-1}; 0.7 μmol L^{-1}) they predicted (a) that the ratio of turnover rate to net flux will be rather large and (b) that near-saturation of adenosine kinase by its substrate adenosine will result in increased sensitivity of metabolic control of cytosolic adenosine. Since at that time neither the concentrations of cytosolic AMP nor that of adenosine were known, this hypothesis could not be tested. In addition the input from the SAH-pathway was not considered.

From the data cited above, it is possible to estimate both the turnover and the net flux through the cycle. Adenosine being formed from (cytosolic + extracellular) AMP and SAH, total adenosine formation being approximately 3.35 nmol min^{-1} g^{-1} and net SAH hydrolysis 0.75 nmol min^{-1} g^{-1}, flux through 5'-nucleotidase can be estimated to be 2.6 nmol min^{-1} g^{-1}. Since in the presence of adenosine deaminase inhibition blocking adenosine kinase increased coronary venous adenosine release from 0.35 ± 0.16 to 3.36 ± 0.71 nmol min^{-1} g^{-1}, the difference (3 nmol min^{-1} g^{-1}) represents an experimental estimate of the flux through adenosine kinase. When taking advantage of a mathematical model[61] that comprehensively describes cardiac adenosine metabolism on the basis of the known kinetic parameters of the relevant enzymes and transport processes, adenosine rephosphorylation was calculated to be 2.6 nmol min^{-1} g^{-1} in the absence of inhibition of either adenosine kinase or adenosine deaminase. Since both AMP dephosphorylation and adenosine phosphorylation in the normoxic guinea pig heart proceed at almost identical rates (2.6 nmol min^{-1} g^{-1}) net flux through this cycle under these conditions is small, while the turnover rate is more than 15-fold greater than net adenosine release from the heart (0.05-0.2 nmol min^{-1} g^{-1}).

What might be the function of this high turnover rate? Assuming a low turnover rate, flux through cytosolic 5'-nucleotidase is small compared to adenosine formation from cytosolic SAH and extracellular AMP. In consequence, changes in cytosolic AMP degradation would have little effect on total adenosine formation, cytosolic adenosine and ultimately adenosine release. However, the major, 10- to 100-fold increase in adenosine release during hypoxia reported by various studies appears to require cytosolic AMP degradation to be the prime source for adenosine, since (a) adenosine formation from SAH has been clearly shown to be independent from cardiac oxygenation[35,70] and (b) adenosine formation from extracellular adenine nucleotides plays only a little role in adenosine release in hypoxia.[18]

This prediction was supported using the comprehensive model of cardiac adenosine metabolism.[61] As can be seen from Figure 2.9, the experimentally-determined turnover rate of the cycle (normal cycling) ensured a rapid, approximately two-fold rise in cytosolic adenosine upon a doubling of hydrolysis of cytosolic AMP. When this turnover rate was reduced to 10% in the model, doubling AMP hydrolysis had hardly any effect on cytosolic adenosine. Conversely, increasing the cycle turnover rate 10-fold did only result in a small additional increase in cytosolic adenosine when compared to the normal situation. Thus, the normal turnover rate of the cycle of 2.6 nmol min^{-1} g^{-1} ensures a close coupling of cytosolic AMP and adenosine in the presence of other sources of adenosine formation (SAH hydrolysis and extracellular AMP). There was, however, no

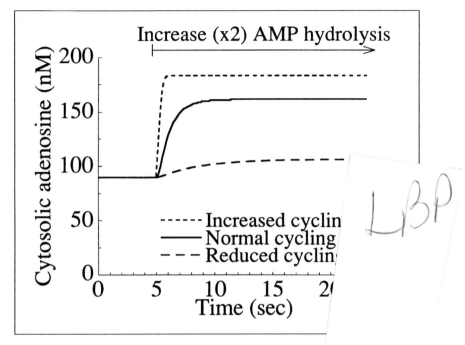

Fig. 2.9. Analysis of the effect of AMP-adenosine cycling on ser
adenosine concentrations to increases in AMP hydrolysis. A n
was employed to predict the effect of doubling the AMP hy
normal conditions of steady-state AMP-adenosine cycling (
and in hypothetical conditions of increased and reduced cycl
tions for the cytosolic adenosine concentration time course   d
with permission from Kroll K et al, Circ Res 1993; 73:846-856.

evidence of adenosine kinase operating near enzyme saturation, nor did the model results indicate that, e.g., a doubling of free AMP induced more than a two-fold rise in cytosolic adenosine.

HYPOXIA-INDUCED INHIBITION OF ADENOSINE KINASE

The model results described above suggested that in the intact heart a doubling of free AMP—e.g., due to a reduction in oxygen supply or a major increase in oxygen demand—would result in a comparable rise in cytosolic adenosine and ultimately adenosine release. This prediction was in line with previous reports of a linear relation between free cytosolic AMP and adenosine release.[20,53] It was, however, at variance with more recent data showing a 10- to 30-fold rise in adenosine release associated with small (2- to 4-fold) changes in free AMP.[28,46,49] The latter data are further supported by a recent study in the saline perfused guinea pig heart[29] demonstrating that global adenosine formation rose proportionally to free AMP in response to hypoxia, whereas adenosine release increased much more. As shown in Figure 2.10 (upper panel) there was a linear correlation between free AMP (200-3000 nmol L^{-1}) and adenosine formation (R^2 = 0.71). While the relative rise in adenosine formation was similar to the relative increase in free AMP (middle panel), the relative rise in adenosine release (lower panel) at each level of oxygen supply was significantly greater. For example, switching from normoxic to hypoxic medium (95 → 40% O_2) increased both free cytosolic AMP and adenosine formation three- to fourfold whereas free cytosolic adenosine and adenosine release rose 15- to 20-fold.

A corollary of the linear relationship between free AMP and adenosine formation is that the hypoxia-induced increase in adenosine formation can be simply explained by mass-action of AMP at cytosolic 5'-nucleotidase. This is consistent with free AMP (ranging from 0.2-3 µmol L^{-1}) being well below K_M of cytosolic 5'-nucleotidase (55-2600 µmol L^{-1}).[24,35,55] Activation of this enzyme as suggested by others[7] is not necessary to explain the observed results.

According to the data shown in Figure 2.10, the rise in adenosine release was much greater than the increase in adenosine formation, pointing to adenosine metabolism as a key factor in the control of cytosolic adenosine. Since the percentage of the adenosine formed that was released increased from 6% (normoxia) to about 20% in the hypoxic heart (40% O_2), 94% of adenosine formed was metabolized in the normoxic heart and less than 80% in hypoxia.[29] This relative decrease in adenosine metabolism could be due to endogenous inhibition of adenosine kinase in the hypoxic heart. Saturation of adenosine kinase at elevated rates of adenosine formation as predicted by Arch and Newsholme[5] might be an alternative explanation. Since V_{max} values of adenosine kinase and adenosine deaminase determined in guinea pig heart extracts (74-228 and 2700 nmol min^{-1} g^{-1}, respectively)[25,35] exceed by far the measured rates of adenosine formation (4-17 nmol min^{-1} g^{-1};

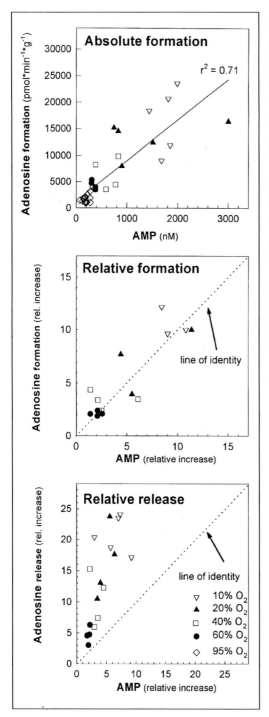

Fig. 2.10. Relation of adenosine formation (measured in the presence of 5 μmol L^{-1} EHNA and 10 μmol L^{-1} iodotubercidin) and coronary venous adenosine release to free cytosolic AMP in the well oxygenated guinea pig heart and at different degrees of hypoxia (60-10% O$_2$). Each heart was exposed to one level of hypoxia only, each symbol represents a single experiment. Free AMP was determined both in the absence and presence of EHNA + iodotubercidin. In the top panel, the measured rate of adenosine formation is related to the concentration of free cytosolic AMP determined. In the middle and bottom panels, adenosine formation and release are expressed relative to the normoxic values of each heart. Reprinted with permission from Decking UKM et al, Circ Res 1997; 81:154-164.

60-10% O_2), enzyme saturation, however, appears highly unlikely. This conclusion was further supported by a model analysis of the experimentally determined data.[29]

Hypoxia-induced inhibition of adenosine kinase in the guinea pig heart was strongly suggested by experiments measuring adenosine release in the presence of selective blockers of adenosine kinase (10 µmol L^{-1} iodotubercidin) or adenosine deaminase (5 µmol L^{-1} EHNA).[29] Blocking adenosine kinase augmented adenosine release several-fold more in the normoxic than in the hypoxic heart, reflecting intrinsic inhibition of adenosine kinase with hypoxia. In contrast, blocking adenosine deaminase increased adenosine release by a similar percentage under both conditions. Mathematical model analysis of these results revealed that the enzyme activity of adenosine kinase (V_{max}/K_M) in hypoxia (40% O_2) was in fact reduced by 94%.[29]

A reduction in adenosine kinase activity associated with an impaired cardiac energy status was also reported in rat cardiomyocytes during ATP depletion (5 mmol L^{-1} iodoacetate).[123] Recent evidence obtained in isolated enzyme studies suggests that one important factor mediating endogenous inhibition of adenosine kinase might be a rise in inorganic phosphate, an obligatory consequence of any decrease in cardiac energy status.[45] Other potential factors include substrate inhibition by adenosine[78] and product inhibition by ADP[94] while lack of ATP substrate is rather unlikely due to the low K_M for ATP (0.1 mmol L^{-1}).[25]

The data now available enable a comprehensive description of flux rates and cytosolic concentrations of the cardiac adenosine metabolism in the normoxic and hypoxic guinea pig heart. As summarized in Figure 2.11, in the normoxic heart >80% of adenosine formed within the heart (from both AMP and SAH) are rephosphorylated to AMP, approximately 15% are deaminated to inosine and less than 5% are released into the coronary venous effluent. In hypoxia (40% O_2), both free cytosolic AMP and the rate of AMP dephosphorylation rise 3- to 4-fold while adenosine formation from SAH remains constant. Flux through adenosine kinase, however, hardly increases due to hypoxia-induced inhibi-

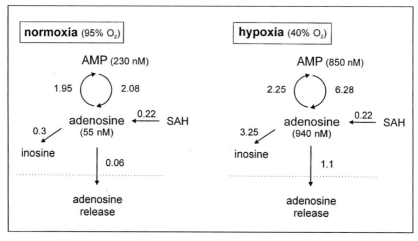

Fig. 2.11. Fluxes and cytosolic concentrations in cardiac adenosine metabolism. Cytosolic concentrations of AMP, adenosine formation (AMP → adenosine + SAH → adenosine) and coronary venous adenosine release rates were experimentally determined as was the relative increase in free cytosolic adenosine (SAH-technique). Basal concentrations of cytosolic adenosine were taken to be 55 nmol L^{-1} based on model predictions and in agreement with previously determined data.[34] Adenosine formation from SAH (transmethylation rate) was estimated on the basis of previously measured data.[35] Flux through adenosine kinase and adenosine deaminase was estimated based on experimental results obtained with selective blockade of adenosine kinase (10 μmol L^{-1} iodotubercidin) and adenosine deaminase (5 μmol L^{-1} EHNA). Adapted with permission from Decking UKM et al, Circ Res 1997; 81:154-164.

tion of the enzyme. As a result cytosolic adenosine rises 17-fold, leading to a similar rise in flux through adenosine deaminase and adenosine release. Thus the hypoxia-induced inhibition of adenosine kinase ultimately translates a small change in cytosolic AMP and adenosine formation into a major rise in cytosolic, interstitial and ultimately vascular adenosine and its metabolites.

Taken together, the AMP-adenosine metabolic cycle couples cytosolic adenosine closely to free cytosolic AMP. The high activity of adenosine kinase facilitates intracellular adenosine (and thus purine) salvage. Since flux through adenosine kinase in the normoxic heart is more than a magnitude higher than net adenosine release, the enzyme is clearly a sensitive control point in the regulation of cytosolic adenosine. This becomes evident, e.g., in hypoxia, where endogenous inhibition of the enzyme amplifies a small impairment of the cardiac energy status into a major rise in cytosolic adenosine, forming the molecular basis for the high sensitivity of the cardiac adenosine system to impaired oxygenation.

AMP DEGRADATION AND CARDIAC ENERGY STATUS

During hypoxia a rise in free AMP induces a proportional increase in adenosine formation, the concomitant endogenous inhibition of adenosine kinase shunts adenosine from purine salvage to release from the heart. This adenosine release makes the cardiac adenylate metabolism an open system which has several important consequences. It is well known that the permeability of the sarcolemmal membrane for ATP, ADP, AMP, creatine and inorganic phosphate is small while that for adenosine and inosine is rather high. Thus any mismatch between ATP synthesis and ATP hydrolysis results in a massive rise in ADP and AMP, enhanced formation of adenosine and ultimately irreversible loss of purines from the heart (see Fig. 2.1). This is consistent with a variety of studies that reported a massive release of purines during low flow ischemia (e.g., refs. 48,52,121) and/or incomplete recovery of ATP during reperfusion.[3,74,87,103,122] This open-system description is at variance with previous, closed-system models of high energy phosphate metabolism during ATP hydrolysis that did not allow for AMP degradation and purine release.[2]

During coronary underperfusion, several studies observed a rapid decrease in PCr associated with a slow decline in ATP. In some of these studies, PCr tended to recover within 30 min to 6 h while ATP continued to decrease.[102,110,111] This was interpreted to indicate that after an initial derangement ATP synthesis exceeded ATP hydrolysis resulting in PCr recovery. However, it is difficult to measure net synthesis and hydrolysis rates of ATP due to its high turnover (20-50 μmol/min/g). Moreover the short half life of the ATP pool (2.5-6 s) suggests that even a small difference between ATP synthesis and demand (e.g., 1%) would result in massive changes in cytosolic ATP within minutes. This, however, had not been observed experimentally.

Recently, a novel explanation has been forwarded.[62] In the saline-perfused rabbit heart, severe underperfusion induced a rapid drop in PCr, followed by a slow decline in ATP and partial PCr recovery (Fig. 2.12, upper panel). The difference between ATP synthesis and hydrolysis was estimated using a mathematical model based on an open-system of myocardial high energy phosphates (Fig. 2.12, lower panel). Assuming a short mismatch between decreased ATP synthesis and continuing consumption, the rapid drop in PCr was accurately predicted. The model accounted for the changes in PCr and ATP only when assuming an ideal match of ATP synthesis and hydrolysis and taking the degradation of AMP to adenosine and inosine into account. Under these conditions the measured purine release was well compatible with the model predictions (Fig. 2.12, middle panel). These data suggest that there is a precise match between ATP synthesis and hydrolysis even during severe underperfusion; AMP hydrolysis may then induce a continuing decrease in ADP and AMP, reflected in a partial recovery of PCr.

Taken together, the enhanced purine release during coronary underperfusion may serve a dual purpose: The enhanced release of adenosine increases interstitial and intravascular adenosine concentrations and activates receptor-mediated transduction pathways. This results in coronary vasodilation, a negative dromotropic and possibly chronotropic effect, opposes the positive inotropic effects of catecholamines and thus acts as a homeostatic metabolite.[105] On the other hand, the continuous degradation of AMP tends to decrease cytosolic AMP and ADP when ATP synthesis and demand are precisely matched thereby improving the cardiac energy status. This latter mechanism therefore contributes to the maintenance of a critical level of free energy of ATP hydrolysis.

CONCLUSIONS AND PERSPECTIVES

One of the most exciting findings of recent adenosine research is that we begin to understand how the formation of this nucleoside is coupled to cardiac energetics. Internal salvage via adenosine kinase not only helps to preserve costly purine compounds but also is responsible for creating an extra- to intracellular adenosine gradient. Adenosine kinase is also target for factors which inhibit the enzyme physiologically and thereby augment adenosine release. P_i

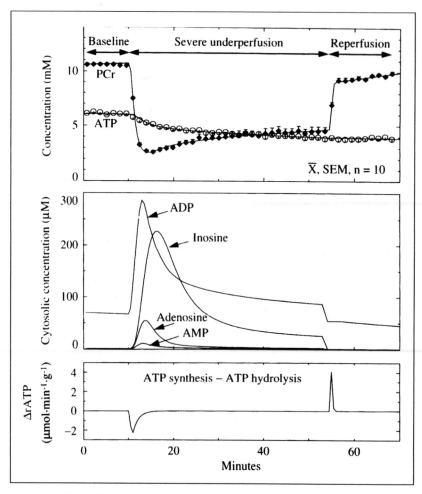

Fig. 2.12. Effect of severe coronary underperfusion (constant 95% flow reduction) in isolated rabbit heart.

Top: NMR data for PCr and ATP. Curves: model fits using the model of an open adenylate system. Fits were obtained by automatically optimizing time-varying function (rATP and the Michaelis constant of 5'-nucleotidase. Intracellular pH and coronary flow were varied according to measurements, all other model parameters were held constant.

Middle: estimates of cytosolic concentrations of ADP, AMP, adenosine, and inosine from model fit in top.

Bottom: estimated time course for (rATP (rate of ATP synthesis minus rate of ATP hydrolysis). For details see text. Reprinted with permission from Kroll K et al, Am J Physiol 1997; 272:H2563-H2576.

is one factor, but other regulatory factors may be still undetected. In view of a high turnover substrate cycle between AMP and adenosine, inhibition of adenosine kinase can potentiate adenosine release with only small changes in the AMP substrate concentration. Due to the effective washout of adenosine, the cardiac adenylate pool is an open system which helps to preserve the free energy charge of ATP hydrolysis. Although most of these concepts have been developed in cardiac tissue, the principal insights discussed in this chapter should be applicable to other organs and tissues as well.

REFERENCES

1. Abraham EH, Prat AG, Gerweck L et al. The multidrug-resistance (mdr1) gene product functions as an ATP channel. Proc Natl Acad Sci USA 1993; 90:312-316.
2. Allen DG, Orchard CH. Myocardial contractile function during ischemia and hypoxia. Circ Res 1987; 60:153-168.
3. Angello DA, Headrick JP, Coddington NM et al. Adenosine antagonism decreases metabolic but not functional recovery from ischemia. Am J Physiol 1991; 260:H193-H200.
4. Arai AE, Pantely GA, Anselone CG et al. Active down-regulation of myocardial energy requirements during prolonged moderate ischemia in swine. Circ Res 1991; 69:1458-1469.
5. Arch JRS, Newsholme EA. Activities and some properties of 5'-nucleotidase, adenosine kinase and adenosine deaminase in tissues from vertebrates and invertebrates in relation to the control of the concentration and the physiological role of adenosine. Biochem J 1978; 174:965-977.
6. Baer HP, Drummond GI, eds. Physiological and Regulatory Functions of Adenosine and Adenine Nucleotides. New York: Raven Press, 1979.
7. Bak MI, Ingwall JS. Acidosis during ischemia promotes adenosine triphosphate resynthesis in postischemic rat heart. In vivo regulation of 5'-nucleotidase. J Clin Invest 1994; 93:40-49.
8. Bardenheuer H, Whelton B, Sparks HV Jr. Adenosine release by the isolated guinea pig heart in response to isoproterenol, acetylcholine, and acidosis: the minimal role of vascular endothelium. Circ Res 1987; 61:594-600.
9. Bassingthwaighte JB, Chaloupka M. Sensivity functions in the estimation of parameters of cellular exchange. Federation Proc 1984; 43:180-184.

10. Bassingthwaighte JB, Goresky CA. Modeling the analysis of solute and water exchange in the microvasculature. In: Renkin EM, Michel CC, eds. Section 2: The Cardiovascular System, Volume IV, Microcirculation, Part I, The American Physiological Society, Bethesda, Maryland, 1984:549-626.

11. Becker BF, Gerlach E. Uric acid, the major catabolite of cardiac adenine nucleotides and adenosine, originates in the coronary endothelium. In: Gerlach E, Becker BF, eds. Topics and Perspectives in Adenosine Research. Berlin: Springer, 1987:209-221.

12. Belardinelli L, Linden J, Berne RM. The cardiac effects of adenosine. Prog Cardiovasc Dis 1989; 32:73-97.

13. Belardinelli L, Pelleg A, eds. Adenosine and Adenine Nucleotides— From Molecular Biology to Integrative Physiology. Boston, USA: Kluwer, 1995.

14. Belardinelli L, Shryock JC, Song Y et al. Ionic basis of the electrophysiological actions of adenosine on cardiomyocytes. FASEB J 1995; 9:359-365.

15. Berne RM. Cardiac nucleotides in hypoxia: possible role in regulation of coronary blood flow. Am J Physiol 1963; 204:317-322.

16. Berne RM. The role of adenosine in the regulation of coronary blood flow. Circ Res 1980; 47:807-813.

17. Berne RM, Rall TW, Rubio R, eds. Regulatory Function of Adenosine. The Hague: Martinus Nijhoff, 1983.

18. Borst M, Schrader J. Adenine nucleotide release from isolated perfused guinea pig hearts and extracellular formation of adenosine. Circ Res 1991; 68:797-806.

19. Borst MM, Deussen A, Schrader J. S-adenosylhomocysteine hydrolase activity in human myocardium. Cardiovasc Res 1992; 26:143-147.

20. Bünger R, Soboll S. Cytosolic adenylates and adenosine release in perfused working heart. Europ J Biochem 1986; 159:203-213.

21. Chagoya de Sanchez V. Circadian variations of adenosine and of its metabolism. Could adenosine be a molecular oscillator for circadian rhythms? Can J Physiol Pharmacol 1995; 73:339-355.

22. Clanachan AS, Paterson ARP, Hammond JR et al. Species differences in nucleoside transport by mammalian erythrocytes. In: Berne RM, Rall TW, Rubio R, eds. Regulatory Functions of Adenosine. Boston: Martinus Nijhoff, 1983:505(Abstr).

23. Cordon-Cardo C, O'Brian JP, Casals D et al. Multidrug-resistance gene (P-glycoprotein) is expressed by endothelial cells at blood-brain barrier sites. Proc Natl Acad Sci USA 1989; 86:695-698.

24. Darvish A, Metting PJ. Purification and regulation of an AMP-specific cytosolic 5'-nucleotidase from dog heart. Am J Physiol 1993; 264:H1528-H1534.

25. De Jong JW. Partial purification and properties of rat-heart adenosine kinase. Arch Int Physiol Biochim 1977; 85:557-569.

26. De Jong JW, Van der Meer P, Nieukoop S et al. Xanthine oxidoreductase activity in perfused hearts of various species, including humans. Circ Res 1990; 67:770-773.

27. De la Haba G, Cantoni GL. The enzymatic synthesis of S-adenosyl-L-homocysteine from adenosine and homocysteine. J Biol Chem 1959; 234:603-608.

28. Decking UKM, Arens S, Schlieper G et al. Dissociation between adenosine release, cardiac energy status and oxygen consumption in working guinea pig hearts. Am J Physiol 1997; 272:H371-H381.

29. Decking UKM, Schlieper G, Kroll K, Schrader J. Hypoxia-induced inhibition of adenosine kinase potentiates cardiac adenosine release. Circ Res 1997; 81:154-164.

30. Dendorfer A, Lauk S, Schaff A et al. New insights into the mechanism of myocardial adenosine formation. In: Gerlach E, Becker BF, eds. Topics and Perspectives in Adenosine Research. Berlin: Springer, 1987:170-184.

31. Des Rosiers C, Nees S. Functional evidence for the presence of adenosine A2-receptors in cultered coronary endothelial cells. Naunyn Schmiedebergs Arch Pharmacol 1987; 336:94-98.

32. Deussen A. Quantitative assessment of sites of adenosine production in the heart. In: Bassingthwaighte JB, Goresky CA, Linehan JH, eds. Whole Organ Approaches to Cellular Metabolism. Capillary Permeation, Cellular Uptake and Product Formation. New York: Springer, 1998.

33. Deussen A, Bading B, Kelm M et al. Formation and salvage of adenosine by macrovascular endothelial cells. Am J Physiol 1993; 264:H692-H700.

34. Deussen A, Borst M, Schrader J. Formation of S-Adenosylhomocysteine in the heart. I: An Index of free intracellular adenosine. Circ Res 1988; 63:240-249.

35. Deussen A, Lloyd HGE, Schrader J. Contribution of S-adenosylhomocysteine to cardiac adenosine formation. J Mol Cell Cardiol 1989; 21:773-782.

36. Deussen A, Möser GH, Schrader J. Contribution of coronary endothelial cells to cardiac adenosine production. Pflügers Arch 1986; 406:608-614.

37. Dobson JG. Mechanism of adenosine inhibition of catecholamine induced responses in heart. Circ Res 1983; 52:151-160.

38. Downey JM, Cohen MV, Ytrehus K et al. Cellular mechanisms in ischemic preconditioning: The role of adenosine and protein kinase C. Ann NY Acad Sci 1994; 723:82-98.

39. Drury AN, Szent-Györgi A. The physiological activity of adenine compounds with especial reference to their action upon the mammalian heart. J Physiol 1929; 68:213-237.

40. Ely SW, Matherne GP, Coleman SD et al. Inhibition of adenosine metabolism increases myocardial interstitial adenosine concentrations and coronary flow. J Mol Cell Cardiol 1992; 24:1321-1332.

41. Gerlach E, Becker BF, eds. Topics and Perspectives in Adenosine Research. Berlin, Heidelberg: Springer, 1987.

42. Gerlach E, Becker BF, Nees S. Formation of adenosine by vascular endothelium. a homeostatic and antithrombogenic mechanism? In: Gerlach E, Becker BF, eds. Topics and Perspectives in Adenosine Research. Berlin: Springer Verlag, 1987:309-319.

43. Gerlach E, Deuticke B, Dreisbach RH. Der Nucleotid-Abbau im Herzmuskel bei Sauerstoffmangel und seine mögliche Bedeutung für die Coronardurchblutung. Die Naturwissenschaften 1963; 50:228-229.

44. Goldman S, Dickinson ES, Slakey LL. Effect of adenosine on synthesis and release of cyclic AMP by cultured vascular cells from swine. J Cyclic Nucl Prot Phosphor Res 1983; 9:69.

45. Gorman MW, He M-X, Hall CS et al. Inorganic phosphate as regulator of adenosine formation in isolated guinea pig hearts. Am J Physiol 1997; 272:H913-H920.

46. Gorman MW, He M-X, Sparks HV. Adenosine formation during hypoxia in isolated hearts: effects of adrenergic blockade. J Mol Cell Cardiol 1994; 26:1613-1623.

47. Griffiths M, Beaumont N, Yao SY et al. Cloning of a human nucleoside transporter implicated in the cellular uptake of adenosine and chemotherapeutic drugs. Nat Med 1997; 3:89-93.

48. He M-X, Gorman MW, Romig GD et al. Adenosine formation and energy status during hypoperfusion and 2-deoxyglucose infusion. Am J Physiol 1991; 260:H917-H926.

49. He M-X, Gorman MW, Romig GD et al. Adenosine formation and myocardial energy status during graded hypoxia. J Mol Cell Cardiol 1992; 24:79-89.

50. He M-X, Wangler RD, Dillon PF et al. Phosphorylation potential and adenosine release during norepinephrine infusion in guinea pig heart. Am J Physiol 1987; 253:H1184-H1191.

51. Headrick JP, Clarke K, Willis RJ. Adenosine production and energy metabolism in ischaemic and metabolically stimulated rat heart. J Mol Cell Cardiol 1989; 21:1089-1100.

52. Headrick JP, Willis RJ. 5'-nucleotidase activity and adenosine formation in stimulated, hypoxic and underperfused rat heart. Biochem J 1989; 261:541-550.

53. Headrick JP, Willis RJ. Adenosine formation and energy metabolism: A ^{31}P-NMR study in isolated rat heart. Am J Physiol 1990; 258:H617-H624.

54. Heusch G, Schulz R. Hibernating myocardium: a review. J Mol Cell Cardiol 1996; 28:2359-2372.

55. Itoh R, Oka J, Ozasa H. Regulation of rat heart cytosol 5'-nucleotidase by adenylate energy charge. Biochem J 1986; 235:847-851.

56. Kobayashi K, Neely JR. Control of maximum rates of glycolysis in rat cardiac muscle. Circ Res 1979; 44:166-175.
57. Kochan Z, Smolenski RT, Yacoub MH et al. Nucleotide and adenosine metabolism in different cell types of human and rat heart. J Mol Cell Cardiol 1994; 26:1497-1503.
58. Kroll K, Bassingthwaighte JB. Role of capillary endothelial cells in transport and metabolism of adenosine in the heart: an example of the impact of endothelial cells on measures of metabolism. In: Bassingthwaighte JB, Goresky CA, Linehan JH, eds. Whole Organ Approaches to Cellular Metabolism, Capillary Permeation, Cellular Uptake and product formation. New York: Springer, 1998.
59. Kroll K, Bukowski TR, Schwartz LM et al. Capillary endothelial transport of uric acid in guinea pig heart. Am J Physiol 1992; 262: H420-H431.
60. Kroll K, Decking UKM, Dreikorn K et al. Rapid turnover of the AMP-adenosine metabolic cycle in the guinea pig heart. Circ Res 1993; 73:846-856.
61. Kroll K, Deussen A, Sweet IR. Comprehensive model of transport and metabolism of adenosine and S-adenosylhomocysteine in the guinea pig heart. Circ Res 1992; 71:590-604.
62. Kroll K, Gustafson LA, Kinzie DJ. Open-system kinetics of myocardial phosphoenergetics during coronary underperfusion. Am J Physiol 1997; 272:H2563-H2576.
63. Kroll K, Martin GV. Steady-state catecholamine stimulation does not increase cytosolic adenosine in canine hearts. Am J Physiol 1994; 265:H503-H510.
64. Kroll K, Schrader J, Piper HM et al. Release of adenosine and cyclic AMP from coronary endothelium in isolated guinea pig hearts: relation to coronary flow. Circ Res 1987; 60:659-665.
65. Kroll K, Stepp DW. Adenosine kinetics in canine coronary circulation. Am J Physiol 1996; 270:H1469-H1483.
66. Lasley RD, Mentzer RM. Protective effects of adenosine in the reversibly injured heart. Ann Thorac Surg 1995; 60:843-846.
67. Lawson JWR, Veech RL. Effects of pH and free Mg2+ on the Keq of creatine kinase reaction and other phosphate hydrolyes and phosphate transfer reactions. J Biol Chem 1979; 254:6528-6537.
68. Linden J. Cloned adenosine A_3 receptors: Pharmacological properties, species differences and receptor functions. Trends Pharmacol Sci 1994; 15:298-306.
69. Lindner F, Rigler R. Über die Beeinflussung der Weite der Herzkranzgefäe durch Produkte des Zellkernstoffwechsels. Pflügers Arch 1931; 226:697-708.
70. Lloyd HGE, Deussen A, Wuppermann H et al. The transmethylation-pathway as a source for adenosine in the isolated guinea-pig heart. Biochem J 1988; 252:489-494.

71. Lloyd HGE, Schrader J. Adenosine metabolism in the guinea pig heart: the role of cytosolic S-adenosyl-L-homocysteine hydrolase, 5'-nucleotidase and adenosine kinase. Eur Heart J 1993; 14(Suppl I):27-33.

72. Loncar R, Flesche CW, Deussen A. Determinants of the S-adenosyl-homocysteine (SAH) technique for the local assessment of cardiac free cytosolic adenosine. J Mol Cell Cardiol 1997; 29:1289-1305.

73. Mallet RT, Hartman DA, Bünger R. Glucose requirement for postischemic recovery of perfused working heart. Eur J Biochem 1990; 188:481-493.

74. Matherne GP, Headrick JP, Berr S et al. Metabolic and functional responses of immature and mature rabbit hearts to hypoperfusion, ischemia, and reperfusion. Am J Physiol Heart Circ Physiol 1993; 264:H2141-H2153.

75. Mattig S, Deussen A. Regulation of vascular adenosine concentration in the guinea pig heart. FASEB J 1997; 11:437(Abstr).

76. Mattig S, Gruber M, Deussen A. Intracellular concentration and transmembraneous gradient of adenosine in macrovascular endothelial cells. Pflügers Arch 1996; Suppl to Vol 431:R128(Abstr).

77. Meghji P, Holmquist CA, Newby AC. Adenosine formation and release from neonatal-rat heart cells in culture. Biochem J 1985; 229:799-805.

78. Miller RL, Adamczyk DL, Miller WH. Adenosine kinase from rabbit liver I. Purification by affinity chromatography and properties. J Biol Chem 1979; 254:2339-2345.

79. Mistry G, Drummond GI. Adenosine metabolism in microvessels from heart and brain. J Mol Cell Cardiol 1986; 18:13-22.

80. Möser GH, Schrader J, Deussen A. Turnover of adenosine in plasma of human and dog blood. Am J Physiol 1989; 256:C799-C806.

81. Mustafa SJ, Marala RB, Abebe W et al. Coronary adenosine receptors: subtypes, localization, and function. In: Belardinelli L, Pelleg A, eds. Adenosine and Adenine Nucleotides: from Molecular Biology to Integrative Physiology. Boston: Kluwer, 1995:229-239.

82. Muxfeldt M, Schaper W. The activity of xanthine oxidase in hearts of pigs, guinea pigs, rabbits, and humans. Basic Res Cardiol 1987; 82:486-492.

83. Nees S, Dendorfer A. New perspectives of myocardial adenine nucleotide metabolism. In: Imai S, Nakazawa M, eds. Role of Adenosine and Adenine Nucleotides in the Biological System. Elsevier, Amsterdam, 1991:273-287.

84. Nees S, Gerlach E. Adenine nucleotide and adenosine metabolism in cultured coronary endothelial cells: Formation and release of adenine compounds and possible functional implications. In: Berne RM, Rall TW, Rubio R, eds. Regulatory Functions of Adenosine. The Hague: Martinus Nijhoff, 1983:347-360.

85. Nees S, Herzog V, Becker BF et al. The coronary endothelium: a highly active metabolic barrier for adenosine. Basic Res Cardiol 1985; 80:515-519.
86. Nees S, Willershausen-Zoennchen B, Gerbes AL et al. Studies on cultured endothelial cells. Folia Angiol 1980; 28:64-68.
87. Neubauer S, Hamman BL, Perry SB et al. Velocity of the creatine kinase reaction decreases in postischemic myocardium: A 31P-NMR magnetization transfer study of the isolated ferret heart. Circ Res 1988; 53:1-15.
88. Newby AC. Adenosine and the concept of "retaliatory metabolites". TIBS 1984; 9:42-44.
89. Newby AC, Holmquist CA. Adenosine production inside rat polymorphonuclear leucocytes. Biochem J 1981; 200:399-403.
90. Newby AC, Worku Y, Meghji P. Critical evaluation of the role of ecto- and cytosolic 5'-nucleotidase in adenosine formation. In: Gerlach E, Becker BF, eds. Topics and Perspectives in Adenosine Research. Berlin: Springer, 1987:155-169.
91. Olah ME, Stiles GL. Adenosine receptor subtypes: Characterization and therapeutic regulation. Annu Rev Pharmacol Toxicol 1995; 35:581-606.
92. Olsson RA, Pearson JD. Cardiovascular Purinoceptors. Physiol Rev 1990; 70:761-845.
93. Olsson RA, Saito D, Steinhart CR. Compartmentalization of adenosine pool of dog and rat hearts. Circ Res 1982; 50:617-626.
94. Palella TD, Andres CM, Fox IH. Human placental adenosine kinase. Kinetic mechanism and inhibition. J Biol Chem 1980; 255:5264-5269.
95. Parkinson FE, Clanachan AS. Adenosine receptors and nucleoside transport sites in cardiac cells. Br J Pharmacol 1991; 104:399-405.
96. Paterson ARP, Jakobs ES, Ng CYC et al. Nucleoside transport inhibition in vitro and in vivo. In: Gerlach E, Becker BF, eds. Topics and Perspectives in Adenosine Research. Berlin: Springer, 1987:89-101.
97. Pearson JD, Carleton JS, Hutchings A et al. Uptake and metabolism of adenosine by pig aortic endothelial and smooth-muscle cells in culture. Biochem J 1978; 170:265-271.
98. Pearson JD, Gordon JL. Vascular endothelium and smooth muscle cells in culture selectively release adenine nucleotides. Nature 1979; 281:384-386.
99. Plagemann PGW, Wohlhueter RM, Woffendin C. Nucleoside and nucleobase transport in animal cells. Biochim Biophys Acta 1988; 947:405-443.
100. Rongen GA, Floras JS, Lenders JW et al. Cardiovascular pharmacology of purines. Clin Sci 1997; 92:13-24.
101. Ross J. Myocardial perfusion-contraction matching. Implications for coronary heart disease and hibernation. Circ 1991; 83:1076-1083.

102. Schaefer S, Carr LJ, Kreutzer U et al. Myocardial adaptation during acute hibernation: Mechanisms of phosphocreatine recovery. Cardiovasc Res 1993; 27:2044-2051.

103. Schaefer S, Prussel E, Carr LJ. Requirement of glycolytic substrate for metabolic recovery during moderate low flow ischemia. J Mol Cell Cardiol 1995; 27:2167-2176.

104. Schrader J. Sites of action and production of adenosine in the heart. In: Burnstock G, ed. Purinergic receptors Receptors and Recognition, Series B, Volume 12. London: Chapman and Hall, 1981:120-162.

105. Schrader J. Adenosine: A homeostatic metabolite in cardiac energy metabolism. Circ 1990; 81:389-391.

106. Schrader J, Baumann G, Gerlach E. Adenosine as inhibitor of myocardial effects of catecholamines. Pflügers Arch 1977; 372:29-35.

107. Schrader J, Gerlach E. Compartmentation of cardiac adenine nucleotides and formation of adenosine. Pflügers Arch 1976; 267:129-135.

108. Schrader J, Haddy FJ, Gerlach E. Release of adenosine, inosine and hypoxanthine from the isolated guinea pig heart during hypoxia, flow-autoregulation and reactive hyperemia. Pflügers Arch 1977; 369:1-6.

109. Schrader J, Schütz W, Bardenheuer H. Role of S-adenosylhomocysteine hydrolase in adenosine metabolism in mammalian heart. Biochem J 1981; 196:65-70.

110. Schulz R, Guth BD, Pieper K et al. Recruitment of an inotropic reserve in moderately ischemic myocardium at the expense of metabolic recovery. A model of short-term hibernation. Circ Res 1992; 70:1282-1295.

111. Schulz R, Rose J, Martin C et al. Development of short-term myocardial hibernation. Its limitation by the severity of ischemia and inotropic stimulation. Circ 1993; 88:684-695.

112. Schütz W, Schrader J, Gerlach E. Different sites of adenosine formation in the heart. Am J Physiol 1981; 240:H963-H970.

113. Shen WK, Kurachi Y. Mechanisms of adenosine-mediated actions on cellular and clinical cardiac electrophysiology. Mayo Clin Proc 1995; 70:274-291.

114. Shoichi I, Nakazawa M, eds. Role of Adenosine and Adenine Nucleotides in the Biological System. Amsterdam: Elsevier Science Publishers BV, 1991.

115. Sonntag M, Deussen A, Schultz J et al. Spatial heterogeneity of blood flow in the dog heart. I. Glucose uptake, free adenosine and oxidative / glycolytic enzyme activity. Pflügers Arch 1996; 432:439-450.

116. Sparks HV Jr, Bardenheuer H. Regulation of adenosine formation by the heart. Circ Res 1986; 58:193-201.

117. Stepp DW, Van Bibber R, Kroll K. Quantitative relation between interstitial adenosine concentration and coronary blood flow. Circ Res 1996; 79:601-610.

118. Sylvén C. Mechanisms of pain in angina pectoris—A critical review of the adenosine hypothesis. Cardiovasc Drugs Ther 1993; 7:745-759.

119. Tavenier M, Skladanowski AC, De Abreu RA et al. Kinetics of adenylate metabolism in human and rat myocardium. Biochim Biophys Acta Gen Subj 1995; 1244:351-356.

120. Thiebaut F, Tsuruo T, Hamada H et al. Cellular localization of the multidrug-resistance gene product P-glycoprotein in normal tissues. Proc Natl Acad Sci USA 1987; 84:7735-7738.

121. Van Belle H, Goosens F, Wynants J. Formation and release of purine catabolites during hypoperfusion, anoxia, and ischemia. Am J Physiol 1987; 252:H886-H893.

122. Vuorinen K, Ylitalo K, Peuhkurinen K et al. Mechanisms of ischemic preconditioning in rat myocardium: Roles of adenosine, cellular energy state, and mitochondrial F_1F_0-ATPase. Circ 1995; 91:2810-2818.

123. Wagner DR, Bontemps F, van den Berghe G. Existence and role of substrate cycling between AMP and adenosine in isolated rabbit cardiomyocytes under control conditions and in ATP depletion. Circ 1994; 90:1343-1349.

124. Wangler RD, Gorman MW, Wang CY et al. Transcapillary adenosine transport and interstitial adenosine concentration in guinea pig hearts. Am J Physiol 1989; 257:H89-106.

CHAPTER 3

Cardiac Electrophysiology of Adenosine: Cellular Basis and Clinical Implications

M.J. Pekka Raatikainen, Donn M. Dennis and Luiz Belardinelli

INTRODUCTION

Adenosine is an endogenous nucleoside that is involved in numerous physiological and pathophysiological processes in mammalian organs and tissues.[1,2] In the heart, adenosine has been shown to increase myocardial oxygen supply through coronary vasodilation and to reduce oxygen demand by decreasing contractility, antagonizing the stimulatory effects of catecholamines, and depressing automaticity and conduction within the sinus and AV nodes.[3-6] In addition, adenosine inhibits platelet aggregation and neutrophil activity, which may prevent myocardial injury due to vascular thrombosis, inhibits generation of oxygen free radicals and the release of cytotoxic enzymes.[7-10] All these potentially cardioprotective effects of adenosine are mediated by specific cell surface membrane receptors.[3,11-14] Currently, four different types of adenosine receptors have been identified, i.e., A_1-, A_{2A}-, A_{2B}- and A_3-adenosine receptors.[12,14,15] Activation of A_1-adenosine receptors mediate both the direct (i.e., negative chronotropic and dromotropic) and indirect (i.e., anti-β-adrenergic) cardiac electrophysiological effects of

Effects of Extracellular Adenosine and ATP on Cardiomyocytes,
edited by Amir Pelleg and Luiz Belardinelli. © 1998 R.G. Landes Company.

adenosine,[11-14,16] and may also be involved in ischemic preconditioning.[17,18] From a clinical point of view, the negative dromotropic effect appears to be the most important electrophysiological action of adenosine. It forms the basis for the efficacy of adenosine in terminating paroxysmal supraventricular tachycardias and for the utility to differentiate various broad and narrow complex tachycardias.[5,19-22]

The objective of this review is to describe the cardiac electrophysiological properties of adenosine. Special emphasis is placed on the cellular electrophysiology, but the therapeutic and diagnostic implications as well as the proarrhythmic effects of adenosine are also described. In addition, we discuss exciting new therapeutic approaches that rely on modulation of the activity and/or concentration of endogenous adenosine.

CARDIAC ELECTROPHYSIOLOGICAL EFFECTS OF ADENOSINE

The cardiac electrophysiological effects of adenosine, regardless of the specific tissue of the heart, are mediated by A_1-adenosine receptors.[5,14,23-25] Like the muscarinic cholinergic receptors and other inhibitory receptors, A_1-adenosine receptors are coupled to its effectors (i.e., ionic channels, adenylate cyclase, and probably protein kinase C and nitric oxide) via a pertussis toxin-sensitive inhibitory G protein(s) (e.g., G_i).[5,14,16,23] The activation of these receptors in the heart can elicit at least two different types of responses: (1) direct (cAMP-independent) effects and (2) indirect (cAMP-dependent or anti-β-adrenergic) effects. In supraventricular tissues adenosine causes both direct and indirect electrophysiological actions, whereas in ventricular myocardium the predominant action of adenosine is to attenuate the effects of catecholamines through inhibition of adenylate cyclase.[5,14,16,23]

Direct (cAMP-independent) effects of adenosine on the electrophysiological properties of specific tissues/cells of the heart are summarized in Table 3.1. These actions are similar to those elicited by acetylcholine, but they are mediated by different receptors.[23] The electrophysiological effects of adenosine, unlike acetylcholine,

Table 3.1. Direct (cAMP-independent) electrophysiological effects of adenosine on specific tissues/cells of the heart

	Membrane Potential	APA	APD	Refractoriness	CV	Excitability
SA node	Hyperpolarization Phase 4 depolarization ↓	↔	↓	↑	N.D.	↓
Atrium	Hyperpolarization Phase 4 depolarization ↓	↔	↓	↓	↔/↑	↓
AV node	Hyperpolarization	↓	↓	↑	↓	↓
Ventricle*	↔	↔	↔	↔	↔	↔/↓

APA, action potential amplitude; APD, action potential duration; CV, conduction velocity. The symbols ↑, ↓, ↔ indicate an increase, a decrease, or no or minor effect, respectively. N.D. = not determined or not reported. *The effects of adenosine on ventricular myocytes are species dependent. For discussion and references see the text.

can be reversed by methylxanthines or adenosine deaminase but not by muscarinic receptor antagonists, e.g., atropine.[3,13] Nucleoside transport blockers (e.g., dipyridamole), inhibitors of adenosine metabolism and allosteric enhancers of A_1-adenosine receptors potentiate the cardiac effects of adenosine but not those of muscarinic agonists.[3,13,26,27] Because adenosine and acetylcholine appear to share the same pertussis toxin-sensitive receptor-effector coupling system,[28] the cardiac actions of both these substances are attenuated in hearts pretreated with pertussis toxin[29] and mimicked by nonhydrolysable GTP analogues.[30] For more detailed reviews on adenosine receptors and receptor-effector coupling mechanisms The reader is referred to previous reviews.[5,11-14,16,23]

SINOATRIAL NODE

The negative chronotropic action of adenosine was first described almost 70 years ago.[31] In their classic paper, Drury and Szent-Györkyi[31] showed that intravenous infusion of adenosine (and AMP) caused sinus slowing in guinea pig, rabbit, dog and cat. Since then numerous subsequent in vivo and in vitro studies have confirmed and expanded these observations.[21,32-37] Although differences in the sensitivity of the sinus node to adenosine have been observed, the negative chronotropic effect of the nucleoside has been found in all species including humans. In humans the mean intravenous dose of adenosine that results in ≥50% increase in sinus cycle length is slightly greater than that required to cause AV block (190 ± 88 versus 179 ± 88 µg/kg).[21] Experimental data indicate the surrounding cells (e.g., crista terminalis) have an apparent lower sensitivity to adenosine than the primary SA nodal pacemaker cells.[35,38] The direct sinus slowing effect of adenosine has been confirmed in experiments using isolated SA node preparations and single SA nodal cells.[32,37] In rabbit isolated SA nodal pacemaker cells, adenosine causes a dose-dependent increase in the maximum diastolic potential and decrease in the rate of depolarization.[32] Because the degree of hyperpolarization in quiescent SA nodal preparations correlates with the degree of sinus slowing,[36] it can be concluded that the slowing of the SA pacemaker activity is due to hyperpolarization of the membrane potential which results in a decrease in phase 4 depolarization.

In addition to the SA node, adenosine also suppresses other cardiac pacemakers, i.e., atrial, AV junctional and ventricular pacemakers.[34,39] Although, as pointed out above, the primary pacemaker cells in sinus node are more sensitive to adenosine than crista terminalis or other surrounding tissues, lower pacemakers appear to have a greater sensitivity to adenosine than the SA node.[25,34,39] Data obtained in several in vitro and in vivo studies have suggested the following sensitivity hierarchy: ventricular pacemaker (His Purkinje) > junctional (AV node) > sinus node.[25,34,39] The mechanism(s) underlying this differential sensitivity remains to be established. One possibility is that different ionic currents play a major role in the mechanism of the negative chronotropic action of adenosine in the nodal tissues and the ventricle.

In patients, bolus injection of adenosine provokes a biphasic response in heart rate. The initial effect, observed within 15-30 seconds of a peripheral venous injection, is sinus bradycardia (i.e., the direct sinus slowing effect) that lasts for approximately 5-10 seconds. This is in general followed by a reflex sinus tachycardia, probably caused by reflex activation of the autonomic nervous system, by direct stimulation of carotid chemoreceptors or by both.[5,25,40] Furthermore, adenosine and vagal stimulation have been suggested to cause complex interactive negative chronotropic effects on the sinus node.[41] In anesthetized dogs heightened background vagal tone significantly augments the action of adenosine, and increased extracellular adenosine concentration potentiates the depressant action of the vagus nerve on the sinus node.[41]

ATRIAL MYOCARDIUM

Experimental studies in isolated atrial myocytes have shown that adenosine causes shortening of action potential duration, hyperpolarization of resting membrane potential and decreases automaticity.[42] As discussed in greater detail below, these effects appear to be due primarily to activation of a specific outward potassium current (I_{KAdo}) that is coupled to A_1-adenosine receptor via a pertussis toxin-sensitive G protein(s). The depressant effect of adenosine on atrial automaticity has also been observed in human atrial fibers obtained from patients undergoing corrective cardiac surgery.[43] In keeping with these findings, it was recently shown in anesthetized dogs[44] and in human patients undergoing routine diagnostic electrophysiological studies,[45] that adenosine causes a dose-dependent shortening of atrial monophasic action potential duration and refractoriness. In both cases these effects were associated with an increased vulnerability to transient atrial arrhythmias.[44,46] Kabell et al[44] also found that the effects of adenosine on atrial monophasic action potential duration were enhanced by the nucleoside transport blocker, dipyridamole and markedly blunted by the adenosine receptor antagonist 8-sulphophenyl-theophylline. In humans, β-adrenergic blockade by propranolol had no effect on the adenosine-induced shortening of atrial monophasic action potential duration and refractoriness.[45]

In a recent clinical study, adenosine caused a slight decrease in atrial conduction times for premature impulses.[47] Although this observation could be explained by the adenosine-induced hyperpolarization of the resting membrane potential, most experimental and clinical studies have not reported changes in atrial conduction velocity. Nevertheless, because of the greater effect of adenosine on refractoriness than on conduction time, the net effect of adenosine should be shortening of atrial wavelength. In general, a shorter wavelength increases the probability of reentrant arrhythmias,[48,49] but as discussed below it may also result in some antiarrhythmic effects during atrial flutter.[44]

ATRIOVENTRICULAR NODE

It is well established that adenosine and A_1-adenosine receptor agonists either prolong AV nodal conduction time or cause higher degree AV block, depending on the concentration of the agonist and rate of atrial pacing.[5,21,50-54] From a clinical point of view, this negative dromotropic effect appears to be the most important electrophysiological action of adenosine. It forms the basis for the efficacy of adenosine in terminating AV nodal reentrant and AV reentrant tachycardias as well as for the value of adenosine in differentiating various broad and narrow complex tachycardias.[5,19-22] Like most cardiac effects of adenosine, the prolongation of AV nodal conduction time was first described by Drury and Szent-Györkyi.[31] Since then the negative dromotropic action of adenosine has been observed in various species and preparations,[21,33,52,55-57] and increasing evidence implicating endogenous adenosine as a biochemical mediator of AV nodal delay during myocardial hypoxia or ischemia has accumulated.[58-62] The depressant effect of adenosine on the AV node is limited to the proximal portion of the AV node, and hence its effect is associated with a prolongation of the atrial-to-His bundle (A-H) conduction time without an effect on the His bundle-to-ventricle (H-V) interval.[21,55] More specifically, the AV node has been shown to consist of at least three different groups of cells, namely the atrionodal (AN), nodal (N) and nodal-His bundle (NH) cells.[21] In guinea pig isolated perfused heart, 83% of the prolongation of the atrioventricular conduction time is due to an increase in the nodal-to-His bundle interval and the remaining 17% due an increase in the atrionodal-to-nodal interval.[55] Clemo and Belardinelli[55] also found that concomitant with an increase in AV conduction delay or block, adenosine concentration dependently decreased the duration and amplitude of the action potential of AN and N cells, and depressed the rate of rise of N cell action potential. Whenever second-degree AV block (or VA block during retrograde conduction) occurred, the N cell action potentials were abolished.[55] In contrast, the action potential of NH cells was not affected by adenosine, nor did adenosine affect the intra-atrial conduction time.[55] Thus, adenosine blocks atrioventricular conduction primarily within the compact N zone of the AV node. Recently, these findings were confirmed in experiments with rabbit isolated single AV nodal cells.[56,57] As in the multicellular preparations, adenosine shortens the duration, depresses the amplitude and reduces the rate of rise of single AV nodal cell action potential. In addition, adenosine causes a significant hyperpolarization of the AV nodal cells.[56,57]

The effect of adenosine on AV nodal conduction time is markedly dependent on the atrial rate. The dose of adenosine required to cause AV block is greater during sinus rhythm than during supraventricular tachycardia.[21] The most probable reason for this rate-dependent action of adenosine in the AV node is that the nucleoside exacerbates the effects of rapid atrial rate on AV nodal conduction delay. The AV node responds to an increase in input rate by prolonging the AV nodal conduction time.[63,64] This is an intrinsic cardioprotective function of the AV node which serves as a filter to protect the heart from life-threatening arrhythmias during episodes of rapid atrial fibrillation and/or atrial flutter. Indeed, it has been shown that adenosine prolongs AV nodal conduction time significantly more at fast than at slow atrial pacing rates.[65-69] In guinea pig and rabbit heart this rate-dependent depression of AV nodal conduction caused by adenosine is associated with increased AV nodal fatigue and reduced facilitation, whereas AV nodal recovery remains unaltered.[66,68] In keeping with these observations, Lai et al[67] have shown that the negative dromotropic effect of adenosine is rate-dependent also in human hearts. The cellular basis of the rate-dependent negative dromotropic effect of adenosine has been studied in rabbit single isolated AV nodal cells. In these cells the rate-dependent activation failure caused by adenosine is accompanied by prolongation of the effective refractory period, an increase in the duration of activation delay and an elevation of the threshold current amplitude required to activate the cells.[57] Thus, the cellular basis for adenosine-induced rate-dependent AV block appears to be the slowed recovery of excitability of AV nodal cells caused by the nucleoside.[57] The augmentation of the negative dromotropic effect of adenosine during fast atrial rates should result in enhanced antiarrhythmic action of adenosine in patients with supraventricular tachycardias.

In contrast to the AV nodal tissues, most accessory AV pathways are insensitive to adenosine.[70,71] The notable exceptions are those AV pathways that exhibit decremental conduction.[72,73] These pathways have similar electrophysiological properties as AV nodal tissue and thus are sensitive to adenosine.[72,73]

VENTRICULAR MYOCARDIUM

The major electrophysiological action of adenosine in ventricles is attenuation of the effects of catecholamines.[3,5,74] It has been observed in all species and it forms the cellular basis of the antiarrhythmic effect of adenosine in patients with cAMP-dependent ventricular tachycardias.[5,23,74-78] In general this effect of adenosine is mediated via inhibition of adenylate cyclase and a decrease in cellular cAMP content,[5,14,23] but intracellular nitric oxide production may also be involved.[79] On the other hand, in ventricular tissues, adenosine has little or no electrophysiological effects under basal conditions except for rat, ferret or other species where demonstrable acetylcholine- and adenosine-activated potassium outward current is present.[80,81] Indeed, adenosine causes no significant effect on resting membrane potential, action potential amplitude, or twitch contraction in ventricular myocytes from guinea pig, rabbit and bovine hearts,[81] whereas in rat and ferret heart ventricular myocytes the nucleoside shortens ventricular action potential duration by 14% and 57%, respectively.[81] In humans adenosine has no significant effect on ventricular monophasic action potential duration.[45] Thus, adenosine should have no effect on reentrant ventricular arrhythmias. However, the effects and the role of adenosine in regulation of ventricular repolarization in the presence of elevated β-adrenergic tone is not well understood. Experimental data from our laboratory indicate that the effects of adenosine on ventricular action potential duration are complex.[82] In guinea pig isolated ventricular myocytes catecholamines may increase or decrease the action potential duration, depending on which ionic current is predominantly affected. Consequently, adenosine, by attenuating the effects of β-adrenergic agonists, either prolongs or shortens the ventricular action potential duration.[82] Although the physiological significance of these observations remains to be determined, they suggest that adenosine may not only suppress but under specific conditions may actually augment the proarrhythmic effects of catecholamines.

As discussed above, adenosine suppresses not only SA node but also all other cardiac pacemakers. Numerous in vivo and in vitro studies have shown that adenosine reduces ventricular automaticity.[39,83-86] Likewise, there is evidence to suggest that the inhibitory effects of hypoxia and ischemia on atrial and ventricular automaticity are mediated by adenosine.[86] Szentmiklosi et al[39] found that ventricular pacemakers are nearly 30-fold more sensitive to the inhibitory effects of adenosine than sinoatrial pacemakers. In agreement with these findings in guinea pig ventricular preparations[39] and those reported for dog Purkinje fibers[85] and rat isolated perfused hearts,[83] adenosine has been shown to depress His Purkinje automaticity in humans.[84] That is, in the presence of isoproterenol adenosine (150-300 µg/kg) decreases the escape rhythm cycle length by approximately 60%. This negative chronotropic effect of adenosine is augmented by dipyridamole and completely abolished by aminophylline, a competitive adenosine receptor antagonist.[84]

In contrast to the above well-recognized observations, it was recently reported that in spontaneously beating isolated right ventricle adenosine at a low concentration range from 0.1-10 nM increased ventricular rate (for a review see ref. 87). However, several pitfalls can be identified which question the relevance of these observations. The concentration of adenosine that was reported to increase ventricular automaticity (0.1-10 nM)[87] is below the normal physiological concentration of the nucleoside. During nonstressed conditions the concentration of adenosine in rat plasma is approximately 80 nM[88] and the interstitial adenosine concentration in rat isolated perfused heart has been estimated to be 60-100 nM.[89,90] Therefore, it is extremely unlikely that adenosine at concentrations several-fold lower than its physiological levels would increase ventricular automaticity. If ventricular automaticity is increased by concentrations of adenosine that are lower (or within) the range of values reported for unstressed cardiac preparations, then adenosine receptor antagonists and adenosine deaminase should slow basal ventricular rate, which is not the case.[83,84,86] In addition, there is a discrepancy between the experimental data of Hernandez and Ribeiro[87] and interpretations of various clinical findings. In rat isolated right ventricular preparations, adenosine was reported to cause "excitatory" actions on ven-

Table 3.2. Effects of adenosine on cardiac membrane ion currents

Ionic Current	Supraventricular Tissues		Ventricular Tissues	
	Direct	Indirect	Direct	Indirect
I_{KAdo}	↑	↔	↔ / ↑*	↔
I_{KATP}	↑	N.D.	↔ / ↑	N.D.
I_K	N.D.	N.D.	↔	↓
I_{CaL}	↓ / ↔	↓	↓	↓
I_F	↓ / ↔	↓		
I_{Cl}	N.D.	↓	↔	↓
I_{Ti}	N.D.	N.D.	↔	↓

Direct effects and indirect effects refer to direct action of adenosine on the ionic current and to an antiadrenergic action mediated by inhibition of adenylate cyclase and reduction in cellular cAMP content, respectively. *Adenosine-sensitive potassium channels are lacking in the ventricular cells of most species. In rat and ferret ventricular myocardium, where adenosine-sensitive potassium channels are present, adenosine activates I_{KAdo}. The symbols ↑, ↓, ↔ indicate an increase, a decrease, or no or minor effect, respectively. N.D. = not determined or not reported. For discussion and references see the text.

tricular pacemakers only at low, subclinical (≤10 nM) concentrations. It is therefore unlikely (if not impossible) that their observations, would explain the occurrence of ventricular premature complexes in patients with acute myocardial ischemia, a condition in which plasma adenosine concentration is elevated to micromolar range.[91,92]

EFFECTS OF ADENOSINE ON MEMBRANE ION CURRENTS

Adenosine is known to modulate the activity of several cardiac membrane ion currents (Table 3.2). Activation of adenosine- (and acetylcholine-) sensitive potassium current (I_{KAdo}) has a paramount role in mediating the direct (cAMP-independent) electrophysiological and inotropic effects of adenosine in supraventricular tissues, but other ion currents such as the pacemaker current (I_F) and the L-type inward calcium current (I_{CaL}) may also contribute to the actions (e.g., modulation of pacemaker activity and rate-dependent depression of AV nodal conduction) of adenosine.[23,82]

On the other hand, by inhibiting the activity of adenylate cyclase adenosine can indirectly antagonize the effects of catecholamines, forskolin, histamine or other agonists on a variety of cAMP-regulated ion currents including L-type calcium current (I_{CaL}), delayed rectifier K^+ current (I_K), chloride currents (I_{Cl}) and pacemaker ionic current (I_F).[23,82]

OUTWARD POTASSIUM CURRENTS

Adenosine Sensitive K^+ Current (I_{KAdo})

The activation of a distinct subset of K^+ channels by adenosine was first described by Belardinelli and Isenberg[42] in isolated guinea pig atrial cells. Subsequently, adenosine has been found to activate an inwardly-rectifying K^+ outward current (I_{KAdo}) (similar to the acetylcholine-regulated K^+ channel, I_{KAch}) also in SA nodal and AV nodal cells.[32,56,57] In contrast, I_{KAdo} is small or absent in most ventricular tissues.[81,82]

In isolated atrial myocytes the shortening of the action potential duration caused by adenosine strongly correlates with an increase in I_{KAdo}, but only minimally with a decrease in basal I_{CaL}.[93] Likewise, it has been found that the reductions in action potential duration and twitch amplitude caused by adenosine are parallel and highly correlated.[94] Furthermore, inhibition of I_{KAdo} caused by intracellular application of cesium chloride and tetraethylammonium reverses the effects of adenosine on both action potential duration and twitch amplitude.[94] In single rabbit SA nodal cells adenosine activates similar time-independent K^+ current as in guinea pig atrial cells, and in SA nodal cell membrane patches adenosine opens single K^+ channels with inward-going rectification and conductances of 25 ± 1.9 pS.[32] Although the maximal amplitude of the K^+ current activated by adenosine is small (30-50 pA), given the large input resistance of the SA nodal cells (~3 GΩ), it should be sufficient to account for the magnitude of the hyperpolarization observed. Thus, direct activation of I_{KAdo} appears to be the principal mechanism whereby adenosine hyperpolarizes atrial and SA nodal cells, reduces the rate of diastolic depolarization and hence suppresses spontaneous pacemaking in sinus node, shortens atrial action potential duration (and refractoriness) and causes direct negative inotropic effects in atria.

Despite the detailed characterization of the depressant effects of adenosine on AV nodal conduction delay and action potential configuration (see above), the mechanism(s) underlying the rate-dependent negative dromotropic effects of adenosine has remained unknown until very recently. Given the important role of activation of I_{KAdo} in mediating the effects of adenosine in SA node and atria,[32,93] it has been suggested that activation of I_{KAdo} would also explain, at least in part, the depressant effects of adenosine in AV node.[55] However, because adenosine not only shortens duration but also reduces amplitude and V_{max} of AV nodal action potential,[57] it is possible that adenosine also affects membrane currents responsible for the rising phase of the action potential (e.g., I_{CaL}). As a result of the recent methodological advances, functionally and morphologically normal cells from the rabbit AV node can now be isolated. Subsequently, Martynyuk et al[56] and Wang et al[57] have demonstrated that adenosine does, indeed, activate I_{KAdo} in rabbit AV nodal cells. However, both groups of investigators also observed a small but significant reduction (30-35%) of basal I_{CaL}.[56,57] Nevertheless, the studies in isolated rabbit AV nodal cells show that activation of I_{KAdo} can in great part explain the negative dromotropic action of adenosine.

In ventricular myocardium, except for rat and ferret, the density of adenosine (and acetylcholine) sensitive K^+ channels is very low,[81,82] which may explain the lack of effect of adenosine on unstimulated action potential duration in most species including humans.[45] In ferret and rat ventricular myocytes, where I_{KAdo} is present, adenosine causes significant shortening of the ventricular action potential duration.[81,82]

ATP Sensitive K^+ Current (I_{KATP})

Although ATP sensitive potassium channels (K_{ATP}) apparently do not operate under normal metabolic conditions, they are rapidly activated when intracellular ATP level falls below a critical concentration.[95] Based on findings that adenosine acts as an endogenous cardioprotective agent,[6,96,97] and probably is a mediator of ischemic preconditioning[17,18] and the in vitro observations of activation of I_{KATP} by adenosine,[98,99] it has been hypothesized that adenosine may activate I_{KATP} in metabolically deprived myocardium. However, before the concept that activation of I_{KATP} is mediated by adenosine, a number of inconsistencies needs to be explained.

1. Adenosine inhibits breakdown of high energy adenine nucleotides during ischemia,[97,100] which in turn should inhibit I_{KATP}.[95]

2. Cyclopentyltheophylline, a specific A_1-adenosine receptor antagonist, reverses hypoxia-induced shortening of atrial, but not ventricular action potential duration in anesthetized guinea pigs.[101]

3. Adenosine, unlike the K_{ATP} channel openers cromakalium and pinacidil, does not shorten ventricular action potential duration.[45,81]

4. Activation of I_{KATP} by dinitrophenol in guinea pig single ventricular myocytes is neither mimicked nor accelerated by either adenosine or adenosine receptor agonists.[82]

5. Adenosine has been reported not to mediate preconditioning in rat heart,[102-104] although it appears to activate I_{KATP} in rat ventricular myocytes.[98]

Furthermore, because glibenclamide, which has been used as specific inhibitor of I_{KATP}, was recently shown to inhibit both the inward rectifier potassium current (I_{K1}) and I_{KAdo},[105] sensitivity to sulfonylureas alone cannot be used to infer participation of I_{KATP} in response to adenosine.

In summary, the role of activation of I_{KATP} in ischemic preconditioning and the mechanisms whereby adenosine could activate I_{KATP} remains to be established. Further studies are needed to address both issues before the physiological role of activation ATP sensitive potassium channels by adenosine in cardiac tissues can be conclusively determined.

Delayed Rectifier K^+ Current (I_K)

The delayed rectifier K^+ current is an important determinant of myocardial repolarization.[106] Adenosine has no effect on nonstimulated I_K, but it markedly attenuates isoproterenol-stimulated I_K.[82] Whether this indirect inhibitory effect of adenosine can modulate ventricular action potential duration in the presence of β-adrenergic stimulation remains to be determined.

L-TYPE CALCIUM INWARD CURRENT (I_{CAL})

The inhibitory effect of adenosine on the catecholamine- or forskolin-stimulated L-type calcium current (I_{CaL}) is well documented.[23,32,74] It is a major effect and may account in great part for the potent anti-β-adrenergic action of adenosine.[23,32,74,107-109] In guinea pig isolated ventricular cells, adenosine decreases the amplitude without altering the voltage dependence or kinetics of activation and inactivation of the isoproterenol-stimulated I_{CaL}.[74] In keeping with these findings, Kato et al[110] later found that in the presence of isoproterenol adenosine decreases the duration of the available state of the calcium channel, without affecting single-channel conductance.

The reduction of basal I_{CaL} and its importance regarding the electrophysiological actions of adenosine in supraventricular tissues has been a matter of extensive study and debate. The major question has been whether adenosine causes a physiologically significant "true" reduction of basal, unstimulated I_{CaL}. In guinea pig atrial myocytes, adenosine reduced basal I_{CaL} by 18-35%.[93,111] However, compared to the 120-300% increase in I_{KAdo}, the inhibitory effect of adenosine on I_{CaL} is minor and apparently contributes little to adenosine-induced shortening of atrial action potential duration and refractoriness.[93,111] In rabbit isolated AV nodal cells adenosine was recently reported to inhibit basal I_{CaL} by approximately 30-35%.[56,57] When assessing the physiological relevance of this observation, it should be noted that, unlike in atrial cells, I_{CaL} is a major contributor of the total inward current in AV nodal cells. Consequently, a small decrease in I_{CaL} is expected to have a much greater impact on the depolarization of AV nodal than atrial myocytes. Nevertheless, these findings need to be confirmed and expanded before any further conclusions can be made regarding the role of inhibition of basal I_{CaL} by adenosine to the well-recognized depressant effect of the nucleoside on AV nodal transmission. In most other species adenosine appears to have little or no significant effect on nonstimulated I_{CaL}.[81,82]

PACEMAKER CURRENT (I_F)

It is well established that adenosine antagonizes the stimulatory effects of catecholamines on the hyperpolarization activated pacemaking current (I_F), and thereby inhibits the positive chronotropic action of β-adrenergic agonists on SA node.[32] Early studies suggested that adenosine (10 μM) has no significant effect on I_F in the absence of isoproterenol.[32] However, more recently, Zaza et al[37] found that adenosine at submicromolar concentrations directly inhibits I_F and slows pacemaking in rabbit sinoatrial myocytes. The mode of inhibition of I_F is similar to that reported for acetylcholine.[37] Of interest, the pacemaker current I_F has also been found in approximately 40% of spontaneously-beating rabbit single AV nodal cardiomyocytes.[57] In keeping with the observations by Zaza et al,[37] adenosine appeared to inhibit nonstimulated I_F in the AV nodal cells.[57]

OTHER IONIC CURRENTS

In addition to modulation of catecholamine-stimulated outward potassium, inward calcium and pacemaker current, adenosine attenuates the effects of β-adrenergic stimulation and forskolin on the transient inward current (I_{Ti}) and the chloride current (I_{Cl}).[82,112] Cellular calcium overload followed by spontaneous oscillatory release of calcium from sarcoplasmic reticulum induces a nonspecific calcium current and/or a Na-Ca exchange current referred to as I_{Ti}. Activation of this transient inward current appears to be associated with delayed afterdepolarization (DAD) and ventricular arrhythmias caused by triggered activity.[112] In single isolated guinea pig myocytes adenosine concomitant with reduction of the amplitude of DADs and abolition of triggered activity suppresses I_{Ti} induced by isoproterenol or forskolin but not the basal I_{Ti} or that induced by ouabain, quinidine or other interventions that do not stimulate adenylate cyclase and raise cellular cAMP.[82,112] Thus, antiarrhythmic effects of adenosine on ventricular arrhythmias caused by cAMP-dependent triggered activity may, at least in part, be explained by suppression of I_{Ti} by the nucleoside.

The cardiac chloride currents (I_{Cl}) are also regulated through a cAMP-dependent pathway. Given the fact that activation of A_1-adenosine receptors effectively inhibits adenylate cyclase, it is likely that adenosine would attenuate the stimulation of I_{Cl} by β-adrenergic agonist. Indeed, adenosine has been found to antagonize stimulation of I_{Cl} by histamine and isoproterenol.[113] The effects of adenosine on other cardiac membrane currents such as inward sodium current (I_{Na}), transient outward current (I_{To}) and T-type calcium current (I_{CaT}) have been studied less extensively. Tytgat et al[114] reported that neither I_{Na} nor I_{CaT} is significantly affected by adenosine. Likewise, we recently demonstrated that adenosine did not significantly affect the transient outward current (I_{To}) either in the presence or in the absence of isoproterenol.[82]

ANTIARRHYTHMIC EFFECTS OF ADENOSINE

As an antiarrhythmic agent adenosine has several unique properties: (1) it is an endogenous substance; (2) it has a very short half-life (≤ 1.5 s); (3) its effects are mediated by specific, G protein-coupled membrane receptors; and (4) it causes markedly different effects in the supraventricular and ventricular myocardium (see refs. 1,3). In addition, due to its rate-dependent action on the AV nodal conduction delay,[65-69] adenosine has only minimal effects during normal sinus rhythm but exerts robust AV nodal depression during supraventricular tachycardias. These properties have not only made adenosine a safe and efficacious drug but they also have set some specific requirements for administration of the drug.[4,5,20,25] The effects of adenosine on specific cardiac arrhythmias are summarized in Table 3.3.

Table 3.3. Effects of adenosine on specific types of cardiac tachyarrhythmias

Tachyarrhythmias that are unaffected (except for provocation of AV block)
- Atrial flutter*
- Atrial fibrillation*
- Atrial tachycardia (some)**
- Reentrant and automatic ventricular tachycardia

Tachyarrhythmias that are suppressed transiently
- Sinus tachycardia
- Atrial tachycardia (some)**

Tachyarrhythmias that terminate
- AV nodal reentrant tachycardia
- AV reentrant tachycardia
- Atrial tachycardia (some)**
- Sinus node reentrant tachycardia
- cAMP-mediated ventricular tachycardia

*Although there are reports of termination of atrial fibrillation and atrial flutter by adenosine, most evidence indicates that rather than terminating, adenosine may actually precipitate atrial flutter and fibrillation. **There is evidence that some atrial tachycardias can be terminated by adenosine, but the precise factors that determine the sensitivity of atrial tachycardias to adenosine remain to be established. For references and further discussion see the text.

SPECIFIC ARRHYTHMIAS

Supraventricular Arrhythmias

Although the negative dromotropic effect of adenosine was first described almost 70 years ago,[31] it was not until the last decade that the clinical relevance of this nucleoside in the treatment of supraventricular tachycardias involving the AV node was fully recognized.[5,20,21,53] The most common supraventricular tachycardia involving the AV node as a part of the reentrant circuit is atrioventricular nodal reentrant tachycardia (AVNRT). In AVNRT the arrhythmogenic substrate is dual pathways within the AV node.[115] During typical AVNRT, the slower pathway maintains antegrade conduction, and retrograde conduction is via a faster pathway. In an atypical form of AVNRT the faster AV nodal pathway conducts in antegrade direction.[115] Another clinically important supraventricular tachycardia involving the AV node is atrioventricular reentrant tachycardia (AVRT) seen in patients with Wolff-Parkinson-White syndrome. In orthodromic AVRT the slower antegrade pathway is within the AV node and the retrograde limb of the reentry circuit is a faster extranodal accessory pathway between the atrium and ventricle, whereas in the less common antidromic AVRT the antegrade conduction is via the accessory pathway.[116] Most evidence indicates that in patients with dual AV nodal pathways, adenosine terminates the tachycardia by abolishing conduction in the slow pathway.[20,21,40,53] However, we recently found in patients with dual AV nodal pathways an excellent correlation between the dose of adenosine that caused AV block and the AV nodal functional refractory period (unpublished data). Furthermore, we were able to induce a greater than 50 ms prolongation of the A-H interval (a "jump" from the fast to the slow pathway) by administrating small doses of adenosine. These data are in agreement with the reported sensitivity order of AV nodal pathways to adenosine: antegrade fast > antegrade slow > retrograde fast pathway.[117] Thus, given the fact that the faster AV nodal pathway has a longer refractory period than the slower pathway, it is possible that, at least in some instances, adenosine may exert its antiarrhythmic action by making the fast pathway completely refractory rather than by blocking conduction in the slow pathway. On the other hand, despite the reported sensitivity of some unusual accessory pathways to adenosine,[72,73] in the vast majority of cases termination of AVRTs by adenosine is due to block of conduction in the AV nodal pathway.

Both retrospective and prospective, double-blind studies have revealed that adenosine is at least as effective as any other antiarrhythmic agent in terminating paroxysmal supraventricular tachycardias with a reported conversion rate of 80-100%.[40,53,118-120] Consistent with the observations that adenosine causes rate-dependent effect on the AV nodal conduction delay,[65-69] the rate of conversion of paroxysmal supraventricular tachycardia to sinus rhythm with adenosine is higher in patients with faster than slower tachycardia.[121] The efficacy of adenosine in pediatric patients[122-124] is very similar to that observed in adults. In addition, preliminary evidence indicates that during pregnancy intravenously administrated adenosine efficaciously terminates maternal[125] and fetal[126] paroxysmal supraventricular tachycardias without causing any major adverse effects. Adenosine appears to have important pharmacological advantages with respect to the other antiarrhythmic agents. The major benefits of adenosine over verapamil include faster onset of action (less than 30 s vs 1-2 minutes) and hemodynamically less severe and shorter-lived side effects.[40] Compared with digitalis glycosides that have been commonly used to treat supraventricular tachycardias in neonates and infants, adenosine appears to be superior in terms of overall efficacy, time to response (less than 30 s vs. up to 8-12 h) and potential for serious adverse effects.[124] With regard to ATP it should be recognized that although adenosine generated by degradation of the nucleotide mediates most of the electrophysiological actions of ATP, the vagomimetic action of ATP may provoke some additional adverse effects.[25]

Adenosine has limited effects on other supraventricular tachycardias. Adenosine causes transient slowing sinus tachycardia and slows the rate or terminates sinus node reentrant tachycardia.[127,128] Early studies suggested that atrial tachycardias are insensitive to adenosine.[129-131] However, more recent data have revealed that some types of atrial arrhythmias may be terminated by adenosine.[127,132] The precise factors that determine whether an atrial arrhythmia is sensitive to adenosine remains to be established. Proposed explanations include that the proximity of the focus of the arrhythmia to the crista terminalis or the mechanism of the arrhythmia determines the sensitivity to adenosine.[127,132] In one study, adenosine terminated 95% of ectopic atrial tachycardias arising from the crista terminalis but only 29% of those arising elsewhere in the atria.[132] On the other hand, the findings by Engelstein et al[127] that adenosine terminated atrial tachycardias due to triggered activity, transiently suppressed automatic atrial tachycardias, and had no effect on reentrant atrial arrhythmias, are consistent with the hypothesis that the response to adenosine is dictated by the mechanism of the tachycardia. The latter observations can be explained on the basis of the cellular electrophysiological effects of adenosine described above. Termination of atrial tachycardias due to triggered activity can be explained by the anti-β-adrenergic effect of the nucleoside and/or direct activation of I_{KAdo}, whereas the transient suppression of automatic atrial tachycardia may be due to hyperpolarization caused by activation of I_{KAdo}. Likewise, the lack of effect on reentrant atrial tachycardias may be explained by the fact that adenosine does not prolong atrial refractoriness. Some experimental data[31,44] and preliminary clinical observation[22] have indicated that adenosine may under certain conditions terminate atrial flutter and fibrillation. However, because the antiarrhythmic effects of adenosine in the setting of atrial flutter and fibrillation have been reported in relatively few experimental protocols, these preliminary observations need to be confirmed in larger clinical trials.

Ventricular Arrhythmias

In lieu of the findings that adenosine has no effects on ventricular repolarization and refractoriness in the absence of catecholamine stimulation, it is not surprising that most forms of ventricular tachycardias appear to be insensitive to adenosine.[22,78,133] Adenosine has no effect on reentrant ventricular tachycardias in which the arrhythmogenic substrate (e.g., myocardial infarction scar, ischemic tissue) is located in the ventricular myocardium or the His-Purkinje system below the His bundle.[22,78,133] Likewise, although adenosine may transiently suppress junctional ectopic tachycardias that originate from the region of the His bundle,[123,134] nearly all automatic ventricular tachycardias are insensitive to adenosine.[22,78,133] However, ventricular tachycardias arising from relatively discrete sites predominantly located in the free wall of the pulmonary infundibulum (i.e., right ventricle outflow tract) can be reliably terminated by adenosine.[61,76-78,135] This arrhythmia is in general characterized by a left bundle branch block and an inferior-axis QRS configuration, and it is most commonly seen in younger patients without structural heart disease.[76,135] Due to its localized nature this tachycardia is often curable by catheter ablation techniques.[135] Although the underlying cellular and electrophysiological mechanisms of this tachycardia have remained somewhat unclear, there is strong experimental and clinical support for the hypothesis that the arrhythmia is due to cAMP-dependent delayed afterdepolarizations and triggered activity.[78,112] The arrhythmia is typically induced by stress, physical exertion or any other condition that is associated with increased catecholamine stimulation.[76,135] The anti-β-adrenergic actions of adenosine are thought to be responsible for termination of such nonreentrant and nonautomatic tachycardias. More specifically, the decrease in adenylate cyclase activity caused by A_1-adenosine receptor activation lowers cellular cAMP which in turn is expected to antagonize ionic currents that are modulated by catecholamines (see above). Because other perturbations that directly or indirectly decrease intracellular calcium (e.g., calcium channel blockers, β-blockers and vagal maneuvers) usually terminate these arrhythmias,[76,78,135] the inhibition of I_{CaL} and I_{Ti} is the most likely mechanism responsible for the effects of adenosine to convert catecholamine-mediated ventricular tachycardias to sinus rhythm.

In experimental animals, adenosine has been shown to reduce the incidence and severity of early ischemic and reperfusion arrhythmias (see refs. 6,97,100). Likewise, adenosine has been postulated to mediate the cardioprotective effects of ischemic precondition at least in some species.[17,18] The cellular and electrophysiological mechanisms of the antiarrhythmic action of adenosine in ischemia/reperfusion arrhythmias and preconditioning are not fully recognized. For example, several seemingly contradictory observations have questioned the activation of I_{KATP} by adenosine (see above). Therefore, despite the preliminary data indicating that adenosine may have beneficial effects on ischemia/reperfusion tachyarrhythmias, further work is needed in this area to elucidate the mechanisms and clinical relevance of adenosine in the setting of ischemia/reperfusion and ischemic preconditioning.

In summary, during recent years adenosine has supplanted verapamil as the drug-of-choice for termination of AV nodal reentrant and AV reentrant tachycardias in hemodynamically stable patients with arrhythmias that are unresponsive to vagotonic physical maneuvers (e.g., Valsalva maneuver, carotid sinus massage). The majority of supraventricular arrhythmias that do not require AV node as part of the arrhythmia circuit are insensitive to adenosine. Ventricular tachycardias, except for the cAMP-mediated ventricular tachycardia, are not affected by adenosine. Finally, it should be emphasized that radiofrequency catheter ablation appears to be a safe and highly effective therapeutic alternative to long-term pharmacotherapy in most if not all adenosine-sensitive tachycardias.

ADENOSINE AS A DIAGNOSTIC TOOL OF CARDIAC ARRHYTHMIAS

In addition to its therapeutic benefits, adenosine has considerable utility for distinguishing the mechanisms of a wide variety of tachycardias (for detailed and elaborate algorithms for the diagnosis of tachycardia with adenosine see refs. 5,20). Therefore, the cardiac rhythm should always be monitored electrocardiographically at least for several minutes following each adenosine bolus, because important diagnostic information can often be obtained at time of the peak effect of adenosine (~30 s after bolus injection) even if tachycardia is not terminated. The rationale for using adenosine as a diagnostic aid is due to its differential effects on the specialized cardiac tissues, short half-life and minimal hemodynamic consequences.[3] In particular, the reproducibility and brevity of AV nodal conduction block elicited by intravenously administrated adenosine has generated interest in using this nucleoside in differentiating ventricular tachycardias from broad complex supraventricular tachycardias and identifying the underlying mechanisms of narrow complex tachycardias. As a general rule, adenosine terminates any tachycardia (either broad or narrow complex) that involves the AV node as a part of its reentrant circuit and has no effect on the majority of other tachycardias of atrial or ventricular origin. However, the prolongation of AV nodal refractoriness and occurrence of transient AV block often reveals the underlying atrial rhythm (e.g., atrial flutter).[5,22,136]

Because of the markedly different action in supraventricular and ventricular myocardium, adenosine administration accurately and safely differentiates ventricular arrhythmias from broad complex supraventricular tachycardias (e.g., supraventricular tachycardia with aberrant conduction or antidromic AVRT).[5,22,136] The reported sensitivity (i.e., revealed the supraventricular origin) and specificity (i.e., had no effect on tachycardia of ventricular origin) for the response to adenosine are 90% and 93%, respectively.[22] The notable exception to this rule is catecholamine (or cAMP)-dependent ventricular tachycardias that originate in the right ventricular outflow tract and terminates following administration of adenosine.[61,76-78,135] Before the availability of adenosine, verapamil was frequently used in the management of wide complex tachycardias thought to be supraventricular in origin.[137] Unlike verapamil, adenosine hardly ever causes serious hemodynamic adverse effects when administered to patients with previously hemodynamically stable ventricular tachycardia or atrial flutter/fibrillation that conducts antegradely over an accessory pathway.[5,22,133,136,137] Initial experience suggests also that adenosine may be used safely in patients with heart failure or taking β-blockers, in whom the use of verapamil in inadvisable.[40]

In narrow complex tachycardia, adenosine is useful in distinguishing between atrial tachycardias from those that involve the AV node as part of the reentrant circuit. In cases in which atrial flutter waves are difficult to discern on a surface electrocardiogram, adenosine-induced transient AV block with perpetuation of the atrial arrhythmia, will markedly facilitate the correct diagnosis and thus guide appropriate therapy.[5,20,22] In a recent study, the accuracy of a group of internal medicine housestaff in diagnosing unknown tachycardias was significantly improved when rhythm strips obtained during peak effect of adenosine were available.[138] For example, the diagnosis of ortodromic reciprocating tachycardia was initially missed by all, but after adenosine administration the arrhythmia was correctly recognized by 45% of the housestaff. In this case, the delta wave and short PR interval became temporarily visible immediately after conversion of the arrhythmia to sinus rhythm. The value of adenosine in diagnosis of intermittent or latent Wolff-Parkinson-White syndrome has been confirmed by others; Garratt and coworkers[139] reported that adenosine given during sinus rhythm was 100% sensitive and specific for latent preexcitation.

Adenosine has also been advocated as a rapid and effective method of assessing the success of radiofrequency or surgical ablation of accessory pathways.[71] The rationale for this diagnostic approach is that adenosine administration effectively depresses (or blocks) conduction through the AV node while having little or no effect on accessory pathway conduction.[70,71] However, because some unusual accessory pathways may be sensitive to adenosine,[72,73,140] evaluation of the response of the accessory pathway to adenosine before radiofrequency ablation is an obligatory part of this diagnostic approach. This technique may emerge particularly valuable for those patients in whom differentiating between retrograde conduction over the AV node or accessory pathway can be difficult, such as patients with posteroseptal or antreoseptal accessory pathway.

In summary, adenosine appears to be a useful, safe and reliable diagnostic tool for differentiating ventricular and supraventricular tachycardias, and in identificating underlying arrhythmogenic mechanism(s) of a variety of tachycardias. In addition, adenosine administration provides a rapid and effective method of assessing the success of radiofrequency or surgical ablation of accessory pathways.

SIDE EFFECTS AND DRUG INTERACTIONS

Because of the ubiquitous nature of adenosine receptors, administration of adenosine commonly causes adverse effects unrelated to the heart. The adverse effects usually occur at the same time as termination of tachycardia and persist for a median of 50 seconds.[53] The reported incidence of adverse symptoms ranges from about 30% to over 70%.[40] The considerable variation in the incidence of side effects is related to how closely the patients are questioned about the symptoms and on the amount of adenosine administrated. It is probable that the majority of patients would experience some minor symptoms if carefully inquired after and if a sufficient dosage is given. The most common patient complaints during adenosine administration are facial flushing, dyspnea and chest discomfort. Other less common side effects include nausea, lightheadedness, headache, dizziness and palpitation.[40,53] In addition, intravenous adenosine administration has occasionally been reported to cause bronchoconstriction particularly in asthmatic patients.[141] Nevertheless, due to the rapid cellular uptake and subsequent metabolism in the vascular endothelial cells, erythrocytes and parenchymal cells, the adverse effects caused by intravenous bolus injections of adenosine are generally brief and well tolerated.

Many of the above described side effects are due to A_2-adenosine receptor-mediated vasodilatory effects of the nucleoside. An example is cutaneous vasodilation which is responsible for the facial flushing. Activation of chemoreceptors by adenosine in the carotid body, on the other hand, may explain the respiratory stimulation and resultant sensation of dyspnea.[75] Chest discomfort or angina-like pain does not appear to be due to adenosine-induced ischemia but rather is due to direct stimulation of cardiac pain receptors by adenosine.[142,143] The potential proarrhythmic actions of adenosine are discussed in detail below.

Interactions with agents that inhibit cellular uptake of adenosine and methylxanthines are of considerable clinical importance in patients receiving adenosine. Nucleoside transport blockers such as dipyridamole competitively inhibit the uptake of adenosine into the cells and prevent subsequent metabolism of the nucleoside. As a result, these agents significantly potentiate and prolong the cardiac actions of adenosine.[27] Clinical doses of dipyridamole potentiates the AV nodal effects of adenosine by a factor of four.[144] Accordingly, it is recommended that the initial adenosine dose given intravenously should not exceed 1 mg in patients receiving dipyridamole.[144] Note that benzodiazepines and calcium channel blockers have also been found to inhibit adenosine uptake, although to a lesser extent than dipyridamole.[6]

Methylxanthines are widely used as antiasthmatic drugs (theophylline) and central nervous stimulants (caffeine). These agents are nonspecific and nonselective (A_1 versus A_2) adenosine receptor antagonists, that may significantly attenuate the effects of adenosine by competitively displacing adenosine from its receptors.[13] Numerous experimental and clinical studies have confirmed that the electrophysiological effects of adenosine are readily reversed by administration of theophylline or other adenosine antagonists.[3,13] Therefore, if administrating adenosine to a patient receiving theophylline (note the risk of bronchoconstriction in asthmatic patients), it should be recognized that larger than usual doses of adenosine may be required. The response to adenosine may also be attenuated in patients who drink large amounts of coffee, tea or other caffeine-containing drinks.

PROARRHYTHMIC EFFECTS OF ADENOSINE

As a result of its potent electrophysiological actions, adenosine not only possesses important antiarrhythmic effects but also some potentially arrhythmogenic actions. However, despite occasional reports of hemodynamically-compromised proarrhythmic episodes, overall adenosine is safe and well-tolerated.

BRADYARRHYTHMIAS

Transient AV nodal block accompanied by a brief sinus bradycardia is commonly seen after intravenous administration of adenosine. In general, the episodes last only a few seconds but prolonged bradycardia, complete heart block and asystole caused by intravenous bolus administration of adenosine have been described.[22,123,134] Patients receiving nucleoside transport blockers (e.g., dipyridamole) are more susceptible to prolonged bradycardia and AV block.[5,40,144] Likewise, there is evidence that patients with acute myocardial infarction, sick sinus syndrome, cardiac arrest or cardiac transplant rejection are at greater risk for the side effects and to the AV blocking effects of adenosine.[145-150]

TACHYARRHYTHMIAS

The potential role of adenosine in facilitating tachyarrhythmias is less well characterized than the nucleotide's negative chronotropic and dromotropic effects. As described above, adenosine, rather than terminating arrhythmias of atrial origin may actually cause some arrhythmogenic actions in the atria.[45,47,53] Similar to acetylcholine,[49] shortening of atrial action potential and refractoriness by adenosine would be expected to produce a decrease in the wavelength (i.e., the product of effective refractory period and conduction velocity) of atrial impulses.[45] Therefore, given the findings that the inducibility of reentrant arrhythmias is directly related to the wavelength,[48,49] it is conceivable that adenosine, by shortening the atrial refractoriness, may facilitate initiation and maintenance of atrial flutter or atrial fibrillation. Indeed, in clinical trials the recommended intravenous dosages of adenosine have been reported to initiate atrial fibrillation in 1-5% of the patients.[40,53] However, it should be noted that the efficacy of adenosine to terminate supraventricular tachycardias involving the AV node is 80-100%.[40,53] Several reasons may account for this observed difference in the anti- and proarrhythmic potential of adenosine. First, adenosine causes hyperpolarization of atrial myocytes.[42] The hyperpolarization of atrial myocytes by adenosine should decrease excitability and possibly lead to an increase in conduction velocity, which in turn would counteract the shortening of atrial wavelength by adenosine. Second, the AV node may be more sensitive to the effects of adenosine than the atria. In keeping with the latter hypothesis, our preliminary results in guinea pig isolated perfused hearts indicated that the concentration-response curve for adenosine to prolong the AV nodal conduction time is significantly shifted to left when the atrial cycle length is shortened from 250-190 ms. In contrast, the concentration-response curves for adenosine to shorten the atrial monophasic action potential duration at 90% repolarization at the slow and fast atrial pacing rates are superimposable.[151] Thus, differences in the rate-dependent effects of adenosine in the AV node and atria may have impact on the relative anti- and proarrhythmic effects of adenosine in the supraventricular tissues.

Atrial arrhythmias such as atrial fibrillation are quite commonly seen during myocardial ischemia. Given the findings that atrial action potential duration and refractoriness shorten[152] and adenosine is released in increased amounts during atrial myocardial hypoxia and ischemia,[60,89] it is tempting to speculate that endogenous adenosine may initiate atrial fibrillation during myocardial ischemia. This interpretation is consistent with the observations that adenosine antagonists reverse the shortening of atrial action potential during hypoxia.[101,152] More recently, atrial fibrillation in two patients that developed the arrhythmia in the early phase of inferior myocardial infarction were converted to normal sinus rhythm within 5 minutes of administration of theophylline.[153] However, despite these interesting preliminary observations, more detailed studies will be needed before it can be widely accepted that elevation of endogenous adenosine concentration may act as a biochemical mediator of atrial flutter and fibrillation in the setting of myocardial ischemia.

The occurrence of ventricular premature complexes (VPC) or short runs of nonsustained ventricular tachycardia upon termination of SVT is relatively common in patients treated with adenosine. For example, the "Adenosine for Paroxysmal Supraventricular Tachycardia Study Group" reported a 33% incidence of VPCs.[53] All episodes were transient and none required interventions. The pathophysiological mechanism(s) underlying VPCs that develop after conversion of supraventricular tachycardia to sinus rhythm by adenosine is not clear. However, a direct proarrhythmic effect seems unlikely for a number of reasons. First, adenosine has no direct effects on membrane ion currents of ventricular myocytes,[5,23,75] and thus it should neither have any antiarrhythmic nor proarrhythmic effect in ventricles. Second, VPCs are not specific for adenosine, but have been observed also after termination of supraventricular tachycardias by verapamil[53] and after surgical ablation of the AV node.[86]

There have been case reports of patients with acquired long QT syndrome who developed polymorphic ventricular tachycardia after receiving adenosine.[154,155] The mechanisms of torsade de pointes are unclear, but most experimental and clinical evidence imply that triggered activity initiated by early afterdepolarizations in the context of prolonged ventricular repolarization provokes the onset of the arrhythmia. Although adenosine has no direct effect on ventricular action potential duration,[45] given the findings that the propensy of generate early afterdepolarizations is enhanced by abrupt slowing of heart rate, conversion of a supraventricular tachycardia to sinus rhythm by adenosine may increase the risk of torsade de pointes in patients with prolonged QT interval. Hence, it is important to recognize long QT syndrome before administration of adenosine or any other agent that promotes bradycardia.

ANTI- AND PROARRHYTHMIC EFFECTS OF ENDOGENOUS ADENOSINE: POTENTIAL THERAPEUTIC IMPLICATIONS

PROARRHYTHMIC EFFECTS OF ENDOGENOUS ADENOSINE

In light of the evidence that endogenous adenosine may act as the biochemical mediator of AV nodal conduction delay and sinus slowing observed during conditions of myocardial oxygen deprivation (see ref. 75 and references therein), it has been proposed that A_1-adenosine receptor antagonists would provide a convenient means to treat sinus bradycardia and/or AV block in the setting of myocardial ischemia/hypoxia.[59,61,156,157] Because the blood supply to the AV node usually arises from the right coronary artery, patients with inferior myocardial infarction are particularly vulnerable to AV nodal blockade. Other clinical conditions in which excessive adenosine production or "supersensitivity" to the nucleoside has been suggested to cause sinoatrial dysfunction or high-degree AV block include postcardiac transplant period,[145-147] sick sinus syndrome[148-150] and postdefibrillation bradyarrhythmias.[158,159]

Aminophylline or theophylline are highly effective in restoring sinus rhythm in patients who developed a clinically significant AV nodal blockade during the early phase of inferior myocardial infarction.[59,156-159] Likewise, sinus node dysfunction occurring early following cardiac transplant and AV nodal block in patients with moderate to severe cardiac allograft rejection respond well to methylxanthines.[145-147,160] Adenosine antagonists are also efficacious in restoring a stable rhythm and blood pressure both in experimental models of ventricular fibrillation/defibrillation and in human patients.[61,161,162] Finally, several independent studies have demonstrated that methylxanthines significantly increase sinus rate, and decrease the incidence and duration of sinus node pauses in patients with sick sinus syndrome.[148-150]

ANTIARRHYTHMIC EFFECTS OF ENDOGENOUS ADENOSINE

Although adenosine is an ideal drug for acute treatment of supraventricular tachycardias, its short half-life makes it unsuitable for chronic management of arrhythmias. In contrast, A_1-receptor agonists with longer half-lives than adenosine have the potential to be useful agents for chronic management of adenosine-sensitive tachycardias. Due to their frequency-dependent negative dromotropic action,[50] A_1-adenosine receptor agonists (like adenosine) have minimal effect during normal heart rates but promote robust depression of AV nodal conduction during tachycardias. However, it is probable that because of their longer half-life, stable A_1-adenosine agonists will cause more severe side effects than the ultra-short acting adenosine.

An alternative and attractive approach which has the potential to mitigate the side effects of adenosine (and receptor agonists) but still retain the nucleoside's potent depressant effects on AV nodal conduction is to modulate the levels and/or activity of endogenous adenosine. Until recently, it was assumed that endogenous adenosine can modulate AV nodal conduction only during myocardial ischemia or hypoxia. However, evidence has now accumulated to demonstrate that pharmacological modulation of endogenous adenosine levels (i.e., inhibition of adenosine removal) and activity (i.e., allosteric enhancement of the A_1-adenosine receptor) can potentially be exploited in the treatment of adenosine-sensitive tachycardias.[26,163-166] Specifically, two different but complementary drug design strategies that elicit event- and site-specific actions have been described (Fig. 3.4).

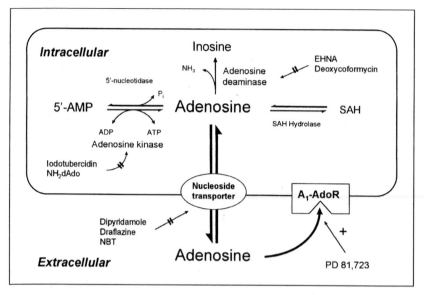

Fig. 3.4. Biochemical pathways of cardiac adenosine metabolism, and the sites of action of various adenosine regulating agents. In the heart, adenosine is formed mainly by dephosphorylation of AMP and to a lesser extend by hydrolysis of S-adenosylhomocysteine to homocysteine and adenosine in reactions catalyzed by 5'-nucleotidase and S-adenosylhomocysteine hydrolase, respectively. A specific nucleoside transporter is responsible for transport of adenosine across the cell membrane. Adenosine taken up by the cells is more likely to be rephosphorylated than deaminated because the affinity of adenosine kinase is much greater than that of adenosine deaminase for adenosine (for more detailed discussion on adenosine metabolism see[14] and references herein). The electrophysiological effects of adenosine are mediated by the A_1-adenosine receptors. Inhibitors of the nucleoside transporter (e.g., dipyridamole; draflazine; and nitrobenzylthioinosine, NBT), adenosine kinase (e.g., idotubercidin; amionodeoxyadenosine, NH_2dAdo) and adenosine deaminase (e.g., erythrononylhydroadenine, EHNA; deoxycoformycin) significantly augment the cardiac effects of adenosine by increasing the interstitial concentration of the nucleoside, whereas PD 81,723 augments the effects of adenosine by allosteric enhancing binding of adenosine to A_1-adenosine receptor.

First, both experimental[26,164] and clinical data[77,144,167] exist to show that it is possible to raise endogenous adenosine to levels sufficient to exert antiarrhythmic effects by using agents that inhibit either nucleoside transporter or enzymes that metabolize adenosine, namely adenosine kinase and adenosine deaminase. In a recent study, nucleoside transport blocker (draflazine) and adenosine kinase inhibitor (idotubercidin, ITU) combined with adenosine deaminase inhibitor (erythro-9-(2-hydroxy-3-nonyl)adenine,

EHNA) significantly elevated epicardial fluid adenosine concentration during fast atrial pacing, whereas during slow pacing they had minimal effect on endogenous adenosine levels.[26] Coincident with the increase in epicardial fluid adenosine concentration, draflazine and ITU+EHNA prolonged the AV nodal conduction time significantly more during fast than slow atrial pacing. The addition of A_1-adenosine receptor antagonist (i.e., cyclopentyltheophylline) or adenosine deaminase abolished the effects of draflazine and ITU+EHNA almost completely.[26] Consistent with these experimental findings, in patients undergoing electrophysiological studies the nucleoside transport blocker, dipyridamole, depresses AV nodal conduction in a frequency-dependent manner via a mechanism involving endogenous adenosine.[144,167] During rapid atrial pacing or supraventricular tachycardia, administration of dipyridamole is associated with marked elevation of coronary sinus adenosine concentration with coincident prolongation of the A-H interval, AV nodal functional refractory period and Wenckebach cycle length.[144,167] In fact, dipyridamole infusion may actually terminate adenosine-sensitive tachycardias including AV nodal reentrant tachycardia, AV reciprocating tachycardia and cAMP-dependent ventricular tachycardia.[77,144,167] In contrast, during normal sinus rhythm or slow atrial pacing dipyridamole causes only minimal effects on AV nodal conduction.[144,167] Like those observed in laboratory animals, the electrophysiological effects of dipyridamole in patients are readily reversed by theophylline, an adenosine antagonist.[144,167]

Second, 2-amino-3-benzoylthiophene derivative (PD 81,723) that allosterically enhances the cardiac effects of adenosine mediated by A_1- but not A_2-adenosine receptors, was recently found to selectively potentiate the cardiac electrophysiological actions of adenosine by increasing the affinity of the A_1-adenosine receptors for adenosine and by promoting receptor-G protein coupling.[26,163-166] Similar to the inhibitors of adenosine metabolism, PD81,723 depresses AV nodal conduction in a rate-dependent manner by a mechanism involving endogenous adenosine acting at the A_1-adenosine receptor.[26] However, the potentiation of

A_1-adenosine receptor-mediated negative dromotropic effect of exogenous adenosine or hypoxia by PD81,723 is not accompanied by any further increase in epicardial transudate or venous effluent concentrations of adenosine.[164,165] This difference in the mechanism of action of PD81,723 and inhibitors of adenosine metabolism is important in light of the results from investigations which have demonstrated that the A_2-adenosine receptor-mediated coronary vasodilation occurs at much lower adenosine concentrations than those required to elicit A_1-adenosine receptor-mediated prolongation of AV nodal conduction time.[96] Unlike PD81,723, which is selective for the A_1-adenosine receptor,[163,164] nucleoside transport blocker and inhibitors of adenosine kinase and adenosine deaminase by relying upon the elevation of endogenous adenosine levels for their effects, are likely to indiscriminately activate all subtypes of adenosine receptors (i.e., A_1, A_{2a}, A_{2b} and A_3) whose affinity for adenosine is within the range of interstitial adenosine concentration achieved. As a consequence, inhibitors of adenosine metabolism may be more likely to cause hemodynamic side effects than selective A_1-adenosine receptor allosteric enhancers.

SUMMARY AND FUTURE DIRECTIONS

In summary, the electrophysiological and pharmacological basis for the safety and efficacy of adenosine in the acute treatment and diagnosis of cardiac arrhythmias are well established. Adenosine has become the drug-of-choice for acute management of paroxysmal supraventricular tachycardias involving the AV node. Likewise, the important role of this nucleoside in the diagnosis of a wide variety of broad and narrow complex tachycardias is also widely accepted. Clinical observations with antiarrhythmic use of theophylline and aminophylline suggest that more specific and selective A_1-adenosine receptor antagonists may be useful in management of cardiac bradyarrhythmias and AV nodal conduction disturbances occurring in settings such as acute myocardial ischemia, cardiac transplant rejection or sick sinus syndrome. On the other hand, despite the promising advances in development of selective and stable adenosine receptor agonists, the use of agents that modulate the levels or activity of endogenous adenosine may prove to be a more attractive approach than direct acting adenosine agonists. These adenosine regulating agents selectively augment the responses to endogenous adenosine or increase the interstitial concentration of adenosine in those organs or specific tissues in which production of the nucleoside is increased. Hence, they should cause less systemic adverse effects than adenosine or direct acting adenosine agonists. The potential of site and event-specific action of adenosine regulating agents has major implications not only in long-term pharmacotherapy of adenosine-sensitive tachyarrhythmias but also in development of new therapies for ischemic heart disease.

REFERENCES

1. Belardinelli L, Pelleg A, eds. Adenosine and Adenine Nucleotides: From Molecular Biology to Integrative Physiology. Boston: Kluwer Academic Press, 1995:1-523.
2. Daval JL, Nehlig A, Nicolas F. Physiological and pharmacological properties of adenosine: therapeutic implications. Life Sci 1991; 49(20):1435-1453.
3. Belardinelli L, Linden J, Berne RM. The cardiac effects of adenosine. Prog Cardiovasc Dis 1989; 32(1):73-97.
4. Faulds D, Chrisp P, Buckley MM. Adenosine. An evaluation of its use in cardiac diagnostic procedures, and in the treatment of paroxysmal supraventricular tachycardia. Drugs 1991; 41(4):596-624.
5. Lerman BB, Belardinelli L. Cardiac electrophysiology of adenosine. Basic and clinical concepts. Circulation 1991; 83(5):1499-1509.
6. Mubagwa K, Mullane K, Flameng W. Role of adenosine in the heart and circulation. Cardiovasc Res 1996; 32(5):797-813.
7. Cristalli G, Camaioni E, Vittori S et al. Selective A_2 adenosine receptor agonists with potent inhibitory activity on platelet aggregation. In: Belardinelli L, Pelleg A, eds. Adenosine and Adenine Nucleotides: From Molecular Biology to Integrative Physiology. Boston: Kluwer Academic Publishers, 1995:141-148.
8. Cronstein BN. Adenosine, an endogenous anti-inflammatory agent. J Appl Physiol 1994; 76(1):5-13.
9. Kitakaze M, Hori M, Sato H et al. Endogenous adenosine inhibits platelet aggregation during myocardial ischemia in dogs. Circ Res 1991; 69(5):1402-1408.
10. Ramkumar V, Nie Z, Rybak LP et al. Adenosine, antioxidant enzymes and cytoprotection. Trends Pharmacol Sci 1995; 16(9):283-285.

11. Kollias-Baker C, Shryock J, Belardinelli L. Myocardial adenosine receptors. In: Belardinelli L, Pelleg A, eds. Adenosine and Adenine Nucleotides: From Molecular Biology to Integrative Physiology. Boston: Kluwer Academic Publishers, 1995:221-228.

12. Olah ME, Stiles GL. Adenosine receptor subtypes: characterization and therapeutic regulation. Annu Rev Pharmacol Toxicol 1995; 35581-35606.

13. Olsson RA, Pearson JD. Cardiovascular purinoceptors. Physiol Rev 1990; 70(3):761-845.

14. Shryock JC, Belardinelli L. Adenosine and adenosine receptors in the cardiovascular system: biochemistry, physiology, and pharmacology. Am J Cardiol 1997; (in press).

15. Tucker AL, Linden J. Cloned receptors and cardiovascular responses to adenosine. Cardiovasc Res 1993; 27(1):62-67.

16. Linden J. Structure and function of A_1 adenosine receptors. FASEB J 1991; 5(12):2668-2676.

17. Downey JM, Cohen MV. Role of adenosine in the phenomenon of ischemic preconditioning. In: Belardinelli L, Pelleg A, eds. Adenosine and Adenine Nucleotides: From Molecular Biology to Integrative Physiology. Boston: Kluwer Academic Publishers, 1995:461-478.

18. Liu GS, Thornton J, Van Winkle DM et al. Protection against infarction afforded by preconditioning is mediated by A_1 adenosine receptors in rabbit heart. Circulation 1991; 84(1):350-356.

19. Belardinelli L, Lerman BB. Electrophysiological basis for the use of adenosine in the diagnosis and treatment of cardiac arrhythmias. Br Heart J 1990; 63(1):3-4.

20. Camm AJ, Garratt CJ. Adenosine and supraventricular tachycardia. N Engl J Med 1991; 325(23):1621-1629.

21. DiMarco JP, Sellers TD, Berne RM et al. Adenosine: electrophysiologic effects and therapeutic use for terminating paroxysmal supraventricular tachycardia. Circulation 1983; 68(6):1254-1263.

22. Rankin AC, Oldroyd KG, Chong E et al. Value and limitations of adenosine in the diagnosis and treatment of narrow and broad complex tachycardias. Br Heart J 1989; 62(3):195-203.

23. Belardinelli L, Shryock JC, Song Y et al. Ionic basis of the electrophysiological actions of adenosine on cardiomyocytes. FASEB J 1995; 9(5):359-365.

24. Belardinelli L, Lerman BB. Adenosine: cardiac electrophysiology. Pacing Clin Electrophysiol 1991; 14(11 Pt 1):1672-1680.

25. Pelleg A, Belardinelli L. Cardiac electrophysiology and pharmacology of adenosine: basic and clinical aspects. Cardiovasc Res 1993; 27(1):54-61.

26. Dennis DM, Raatikainen MJP, Martens JR et al. Modulation of atrioventricular nodal function by metabolic and allosteric regulators of endogenous adenosine in guinea pig heart. Circulation 1996; 94(10):2551-2559.

27. Van Belle H. Nucleoside transport inhibition: a therapeutic approach to cardioprotection via adenosine? Cardiovasc Res 1993; 27(1):68-76.
28. Böhm M, Gierschik P, Schwinger RH et al. Coupling of M-cholinoceptors and A_1 adenosine receptors in human myocardium. Am J Physiol 1994; 266(5 Pt 2):H1951-H1958.
29. Böhm M, Schmitz W, Scholz H et al. Pertussis toxin prevents adenosine receptor- and m-cholinoceptor-mediated sinus rate slowing and AV conduction block in the guinea-pig heart. Naunyn Schmiedebergs Arch Pharmacol 1989; 339(1-2):152-158.
30. Yatani A, Codina J, Brown AM et al. Direct activation of mammalian atrial muscarinic potassium channels by GTP regulatory protein Gk. Science 1987; 235(4785):207-211.
31. Drury AN, Szent-Györkyi A. The physiological activity of adenine compounds with special reference to their action upon the mammalian heart. J Physiol (Lond) 1926; 68213-237.
32. Belardinelli L, Giles WR, West A. Ionic mechanisms of adenosine actions in pacemaker cells from rabbit heart. J Physiol (Lond) 1988; 405615-633.
33. Froldi G, Belardinelli L. Species-dependent effects of adenosine on heart rate and atrioventricular nodal conduction. Mechanism and physiological implications. Circ Res 1990; 67(4):960-978.
34. Pelleg A, Hurt C, Miyagawa A et al. Differential sensitivity of cardiac pacemakers to exogenous adenosine in vivo. Am J Physiol 1990; 258(6 Pt 2):H1815-H1822.
35. West GA. Actions of adenosine on the sinus node. Prog Clin Biol Res 1987; 23097-108.
36. West GA, Belardinelli L. Correlation of sinus slowing and hyperpolarization caused by adenosine in sinus node. Pflügers Arch 1985; 403(1):75-81.
37. Zaza A, Rocchetti M, DiFrancesco D. Modulation of the hyperpolarization-activated current (I_f) by adenosine in rabbit sinoatrial myocytes. Circulation 1996; 94(4):734-741.
38. West GA, Belardinelli L. Sinus slowing and pacemaker shift caused by adenosine in rabbit SA node. Pflügers Arch 1985; 403(1):66-74.
39. Szentmiklosi AJ, Nemeth M, Szegi J et al. Effect of adenosine on sinoatrial and ventricular automaticity of the guinea pig. Naunyn Schmiedebergs Arch Pharmacol 1980; 311(2):147-149.
40. Rankin AC, Brooks R, Ruskin JN et al. Adenosine and the treatment of supraventricular tachycardia. Am J Med 1992; 92(6):655-664.
41. Pelleg A, Mitsuoka T, Mazgalev T et al. Interacting negative chronotropic effects of adenosine and the vagus nerve on the canine sinus node. Cardiovasc Res 1988; 22(1):55-61.
42. Belardinelli L, Isenberg G. Isolated atrial myocytes: adenosine and acetylcholine increase potassium conductance. Am J Physiol 1983; 244(5):H734-H737.

43. Wang GY, Wei J, Lin CI. Depressant effect of adenosine in isolated human and canine atrial fibres. Cardiovasc Res 1991; 25(1):31-35.
44. Kabell G, Buchanan LV, Gibson JK et al. Effects of adenosine on atrial refractoriness and arrhythmias. Cardiovasc Res 1994; 28(9): 1385-1389.
45. O'Nunain S, Garratt C, Paul V et al. Effect of intravenous adenosine on human atrial and ventricular repolarisation. Cardiovasc Res 1992; 26(10):939-943.
46. O'Nunain S, Jennison S, Bashir Y et al. Effects of adenosine on atrial repolarization in the transplanted human heart. Am J Cardiol 1993; 71(2):248-251.
47. Kabell G, Karas BJ, Corbisiero R et al. Effects of adenosine on wavelength of premature atrial complexes in patients without structural heart disease. Am J Cardiol 1996; 78(12):1443-1446.
48. Janse MJ. What a clinician needs to know about the mechanisms of action of antiarrhythmic drugs. Clin Cardiol 1991; 14(1):65-67.
49. Rensma PL, Allessie MA, Lammers WJ et al. Length of excitation wave and susceptibility to reentrant atrial arrhythmias in normal conscious dogs. Circ Res 1988; 62(2):395-410.
50. Belardinelli L, Lu J, Dennis D et al. The cardiac effects of a novel A_1-adenosine receptor agonist in guinea pig isolated heart. J Pharmacol Exp Ther 1994; 271(3):1371-1382.
51. Belardinelli L. Modulation of atrioventricular transmission by adenosine. Prog Clin Biol Res 1987; 230:109-118.
52. Belardinelli L, Shryock J, West GA et al. Effects of adenosine and adenine nucleotides on the atrioventricular node of isolated guinea pig hearts. Circulation 1984; 70(6):1083-1091.
53. DiMarco JP, Miles W, Akhtar M et al. Adenosine for paroxysmal supraventricular tachycardia: dose ranging and comparison with verapamil. Assessment in placebo- controlled, multicenter trials. The Adenosine for PSVT Study Group. Ann Intern Med 1990; 113(2): 104-110.
54. Garratt CJ, Malcolm AD, Camm AJ. Adenosine and cardiac arrhythmias. BMJ 1992; 305(6844):3-4.
55. Clemo HF, Belardinelli L. Effect of adenosine on atrioventricular conduction. I: Site and characterization of adenosine action in the guinea pig atrioventricular node. Circ Res 1986; 59(4):427-436.
56. Martynyuk AE, Kane KA, Cobbe SM et al. Adenosine increases potassium conductance in isolated rabbit atrioventricular nodal myocytes. Cardiovasc Res 1995; 30(5):668-675.
57. Wang D, Shryock JC, Belardinelli L. Cellular basis for the negative dromotropic effect of adenosine on rabbit single atrioventricular nodal cells. Circ Res 1996; 78(4):697-706.
58. Belardinelli L, Belloni FL, Rubio R et al. Atrioventricular conduction disturbances during hypoxia. Possible role of adenosine in rabbit and guinea pig heart. Circ Res 1980; 47(5):684-691.

59. Bertolet BD, McMurtrie EB, Hill JA et al. Theophylline for the treatment of atrioventricular block after myocardial infarction. Ann Intern Med 1995; 123(7):509-511.

60. Clemo HF, Belardinelli L. Effect of adenosine on atrioventricular conduction. II: Modulation of atrioventricular node transmission by adenosine in hypoxic isolated guinea pig hearts. Circ Res 1986; 59(4):437-446.

61. Wesley RC, Jr., Lerman BB, DiMarco JP et al. Mechanism of atropine-resistant atrioventricular block during inferior myocardial infarction: possible role of adenosine. J Am Coll Cardiol 1986; 8(5):1232-1234.

62. Xu J, Tong H, Wang L et al. Endogenous adenosine, A_1 adenosine receptor, and pertussis toxin sensitive guanine nucleotide binding protein mediate hypoxia induced AV nodal conduction block in guinea pig heart in vivo. Cardiovasc Res 1993; 27(1):134-140.

63. Billette J, Nattel S. Dynamic behavior of the atrioventricular node: a functional model of interaction between recovery, facilitation, and fatigue. J Cardiovasc Electrophysiol 1994; 5(1):90-102.

64. Meijler FL, Janse MJ. Morphology and electrophysiology of the mammalian atrioventricular node. Physiol Rev 1988; 68(2):608-647.

65. Dennis D, Jacobson K, Belardinelli L. Evidence of spare A_1-adenosine receptors in guinea pig atrioventricular node. Am J Physiol 1992; 262(3 Pt 2):H661-H671.

66. Jenkins JR, Belardinelli L. Atrioventricular nodal accommodation in isolated guinea pig hearts: physiological significance and role of adenosine. Circ Res 1988; 63(1):97-116.

67. Lai WT, Lee CS, Wu SN. Rate-dependent properties of adenosine-induced negative dromotropism in humans. Circulation 1994; 90(4):1832-1839.

68. Nayebpour M, Billette J, Amellal F et al. Effects of adenosine on rate-dependent atrioventricular nodal function. Potential roles in tachycardia termination and physiological regulation. Circulation 1993; 88(6):2632-2645.

69. Stark G, Sterz F, Stark U et al. Frequency-dependent effects of adenosine and verapamil on atrioventricular conduction of isolated guinea pig hearts. J Cardiovasc Pharmacol 1993; 21(6):955-959.

70. Garratt CJ, Griffith MJ, O'Nunain S et al. Effects of intravenous adenosine on antegrade refractoriness of accessory atrioventricular connections. Circulation 1991; 84(5):1962-1968.

71. Keim S, Curtis AB, Belardinelli L et al. Adenosine-induced atrioventricular block: a rapid and reliable method to assess surgical and radiofrequency catheter ablation of accessory atrioventricular pathways. J Am Coll Cardiol 1992; 19(5):1005-1012.

72. Lerman BB, Greenberg M, Overholt ED et al. Differential electrophysiologic properties of decremental retrograde pathways in long RP' tachycardia. Circulation 1987; 76(1):21-31.

73. Engelstein ED, Wilber D, Wadas M et al. Limitations of adenosine in assessing the efficacy of radiofrequency catheter ablation of accessory pathways. Am J Cardiol 1994; 73(11):774-779.

74. Isenberg G, Belardinelli L. Ionic basis for the antagonism between adenosine and isoproterenol on isolated mammalian ventricular myocytes. Circ Res 1984; 55(3):309-325.

75. Belardinelli L, Curtis AB, Bertolet BD. Cardiac electrophysiology of adenosine: antiarrhythmic and proarrhythmic actions. In: Jacobson KA, Jarvis MF, eds. Purinergic Approaches in Experimental Therapeutics. New York: Wiley-Liss, Inc, 1997:185-202.

76. Lerman BB, Stein KM, Markowitz SM. Adenosine-sensitive ventricular tachycardia: a conceptual approach. J Cardiovasc Electrophysiol 1996; 7(6):559-569.

77. Lerman BB. Response of nonreentrant catecholamine-mediated ventricular tachycardia to endogenous adenosine and acetylcholine. Evidence for myocardial receptor-mediated effects. Circulation 1993; 87(2):382-390.

78. Lerman BB, Belardinelli L, West GA et al. Adenosine-sensitive ventricular tachycardia: evidence suggesting cyclic AMP-mediated triggered activity. Circulation 1986; 74(2):270-280.

79. Martynyuk AE, Kane KA, Cobbe SM et al. Nitric oxide mediates the anti-adrenergic effect of adenosine on calcium current in isolated rabbit atrioventricular nodal cells. Pflügers Arch 1996; 431(3):452-457.

80. Belardinelli L, Isenberg G. Actions of adenosine and isoproterenol on isolated mammalian ventricular myocytes. Circ Res 1983; 53(3):287-297.

81. Song Y, Belardinelli L. Electrophysiological and functional effects of adenosine on ventricular myocytes of various mammalian species. Am J Physiol 1996; 271(4 Pt 1):C1233-C1243.

82. Song Y, Shryock J, Belardinelli L. Modulation of cardiomyocyte membrane currents by A_1 adenosine receptors. In: Belardinelli L, Pelleg A, eds. Adenosine and Adenine Nucleotides: From Molecular Biology to Integrative Physiology. Boston: Kluwer Academic Publishers, 1995:97-102.

83. Heller LJ, Olsson RA. Inhibition of rat ventricular automaticity by adenosine. Am J Physiol 1985; 248(6 Pt 2):H907-H913.

84. Lerman BB, Wesley RC, Jr., DiMarco JP et al. Antiadrenergic effects of adenosine on His-Purkinje automaticity. Evidence for accentuated antagonism. J Clin Invest 1988; 82(6):2127-2135.

85. Rosen MR, Danilo P, Jr., Weiss RM. Actions of adenosine on normal and abnormal impulse initiation in canine ventricle. Am J Physiol 1983; 244(5):H715-H721.

86. Wesley RC, Jr., Belardinelli L. Role of adenosine on ventricular overdrive suppression in isolated guinea pig hearts and Purkinje fibers. Circ Res 1985; 57(4):517-531.

87. Hernandez J, Ribeiro JA. Excitatory actions of adenosine on ventricular automaticity. Trends Pharmacol Sci 1996; 17(4):141-144.
88. Phillis JW, O'Regan MH, Perkins LM. Measurement of rat plasma adenosine levels during normoxia and hypoxia. Life Sci 1992; 51(15):PL149-52.
89. Raatikainen MJ, Peuhkurinen KJ, Hassinen IE. Contribution of endothelium and cardiomyocytes to hypoxia-induced adenosine release. J Mol Cell Cardiol 1994; 26(8):1069-1080.
90. Raatikainen MJP, Peuhkurinen KJ, Hassinen IE. Cellular source and role of adenosine in isoproterenol-induced coronary vasodilatation. J Mol Cell Cardiol 1991; 23(10):1137-1148.
91. Lazzarino G, Raatikainen P, Nuutinen M et al. Myocardial release of malondialdehyde and purine compounds during coronary bypass surgery. Circulation 1994; 90(1):291-297.
92. Nissinen J, Raatikainen MJ, Karlqvist K et al. Efflux of adenosine and its catabolites during cold blood cardioplegia. Ann Thorac Surg 1993; 55(6):1546-1552.
93. Visentin S, Wu SN, Belardinelli L. Adenosine-induced changes in atrial action potential: contribution of Ca and K currents. Am J Physiol 1990; 258(4 Pt 2):H1070-H1078.
94. Wang D, Belardinelli L. Mechanism of the negative inotropic effect of adenosine in guinea pig atrial myocytes. Am J Physiol 1994; 267(6 Pt 2):H2420-H2429.
95. Wilde AA, Janse MJ. Electrophysiological effects of ATP sensitive potassium channel modulation: implications for arrhythmogenesis. Cardiovasc Res 1994; 28(1):16-24.
96. Belardinelli L, Shryock J. Does adenosine function as a retaliatory metabolite in the heart? News in Physiological Sciences 1992; 7:52-56.
97. Ely SW, Berne RM. Protective effects of adenosine in myocardial ischemia. Circulation 1992; 85(3):893-904.
98. Kirsch GE, Codina J, Birnbaumer L et al. Coupling of ATP-sensitive K^+ channels to A_1 receptors by G proteins in rat ventricular myocytes. Am J Physiol 1990; 259(3 Pt 2):H820-H826.
99. Li GR, Feng J, Shrier A et al. Contribution of ATP-sensitive potassium channels to the electrophysiological effects of adenosine in guinea-pig atrial cells. J Physiol (Lond) 1995; 484(Pt 3):629-642.
100. Lasley RD, B2nger R, Mentzer RM. Receptor-mediated and metabolic effects of adenosine in ischemic and postischemic myocardium. In: Belardinelli L, Pelleg A, eds. Adenosine and Adenine Nucleotides: From Molecular Biology to Integrative Physiology. Boston: Kluwer Academic Publishers, 1995:351-360.
101. Xu J, Wang L, Hurt CM et al. Endogenous adenosine does not activate ATP-sensitive potassium channels in the hypoxic guinea pig ventricle in vivo. Circulation 1994; 89(3):1209-1216.

102. Asimakis GK, Inners-McBride K, Conti VR. Attenuation of post-ischaemic dysfunction by ischaemic preconditioning is not mediated by adenosine in the isolated rat heart. Cardiovasc Res 1993; 27(8):1522-1530.

103. Li Y, Kloner RA. The cardioprotective effects of ischemic 'preconditioning' are not mediated by adenosine receptors in rat hearts. Circulation 1993; 87(5):1642-1648.

104. Vuorinen K, Ylitalo K, Peuhkurinen K et al. Mechanisms of ischemic preconditioning in rat myocardium. Roles of adenosine, cellular energy state, and mitochondrial F1F0- ATPase. Circulation 1995; 91(11):2810-2818.

105. Song Y, Srinivas M, Belardinelli L. Nonspecific inhibition of adenosine-activated K+ current by glibenclamide in guinea pig atrial myocytes. Am J Physiol 1996; 271(6 Pt 2):H2430-H2437.

106. Barry DM, Nerbonne JM. Myocardial potassium channels: electrophysiological and molecular diversity. Annu Rev Physiol 1996; 58:363-394.

107. Belardinelli L, Vogel S, Linden J et al. Antiadrenergic action of adenosine on ventricular myocardium in embryonic chick hearts. J Mol Cell Cardiol 1982; 14(5):291-294.

108. Dobson JG, Jr., Fenton RA, Romano FD. The cardiac anti-adrenergic effect of adenosine. Prog Clin Biol Res 1987; 230:331-343.

109. Schrader J, Baumann G, Gerlach E. Adenosine as inhibitor of myocardial effects of catecholamines. Pflügers Arch 1977; 372(1):29-35.

110. Kato M, Yamaguchi H, Ochi R. Mechanism of adenosine-induced inhibition of calcium current in guinea pig ventricular cells. Circ Res 1990; 67(5):1134-1141.

111. Cerbai E, Klockner U, Isenberg G. Ca-antagonistic effects of adenosine in guinea pig atrial cells. Am J Physiol 1988; 255(4 Pt 2): H872-H878.

112. Song Y, Thedford S, Lerman BB et al. Adenosine-sensitive afterdepolarizations and triggered activity in guinea pig ventricular myocytes. Circ Res 1992; 70(4):743-753.

113. Harvey RD, Hume JR. Histamine activates the chloride current in cardiac ventricular myocytes. J Cardiovasc Electrophysiol 1990; 1:309-317.

114. Tytgat J, Nilius B, Vereecke J et al. The T-type Ca channel in guinea-pig ventricular myocytes is insensitive to isoproterenol. Pflügers Arch 1988; 411(6):704-706.

115. Akhtar M, Jazayeri MR, Sra J et al. Atrioventricular nodal reentry. Clinical, electrophysiological, and therapeutic considerations. Circulation 1993; 88(1):282-295.

116. Narula OS. Wolff-Parkinson-White Syndrome. A review. Circulation 1973; 47(4):872-887.

117. Lai WT, Wu JC, Hwang YS et al. Differential effect of adenosine on antegrade fast, antegrade slow and retrograde fast pathways in patients with atrioventricular nodal reentrant tachycardia. Circulation 1996; 94(8):I-284(Abstract).

118. Garratt C, Linker N, Griffith M et al. Comparison of adenosine and verapamil for termination of paroxysmal junctional tachycardia. Am J Cardiol 1989; 64(19):1310-1316.

119. Sellers TD, Kirchhoffer JB, Modesto TA. Adenosine: a clinical experience and comparison with verapamil for the termination of supraventricular tachycardias. Prog Clin Biol Res 1987; 230:283-299.

120. Rankin AC, Oldroyd KG, Chong E et al. Adenosine or adenosine triphosphate for supraventricular tachycardias? Comparative double-blind randomized study in patients with spontaneous or inducible arrhythmias. Am Heart J 1990; 119(2 Pt 1):316-323.

121. Gausche M, Persse DE, Sugarman T et al. Adenosine for the prehospital treatment of paroxysmal supraventricular tachycardia. Ann Emerg Med 1994; 24(2):183-189.

122. Crosson JE, Etheridge SP, Milstein S et al. Therapeutic and diagnostic utility of adenosine during tachycardia evaluation in children. Am J Cardiol 1994; 74(2):155-160.

123. Till J, Shinebourne EA, Rigby ML et al. Efficacy and safety of adenosine in the treatment of supraventricular tachycardia in infants and children. Br Heart J 1989; 62(3):204-211.

124. Clarke B, Till J, Rowland E et al. Rapid and safe termination of supraventricular tachycardia in children by adenosine. Lancet 1987; 1(8528):299-301.

125. Blanch G, Walkinshaw SA, Walsh K. Cardioversion of fetal tachyarrhythmia with adenosine. Lancet 1994; 344(8937):1646.

126. Elkayam U, Goodwin TM. Adenosine therapy for supraventricular tachycardia during pregnancy. Am J Cardiol 1995; 75(7):521-523.

127. Engelstein ED, Lippman N, Stein KM et al. Mechanism-specific effects of adenosine on atrial tachycardia. Circulation 1994; 89(6):2645-2654.

128. Griffith MJ, Garratt CJ, Ward DE et al. The effects of adenosine on sinus node reentrant tachycardia. Clin Cardiol 1989; 12(7):409-411.

129. DiMarco JP, Sellers TD, Lerman BB et al. Diagnostic and therapeutic use of adenosine in patients with supraventricular tachyarrhythmias. J Am Coll Cardiol 1985; 6(2):417-425.

130. Epstein ML, Belardinelli L. Failure of adenosine to terminate focal atrial tachycardia. Pediatr Cardiol 1993; 14(2):119-121.

131. Haines DE, DiMarco JP. Sustained intraatrial reentrant tachycardia: clinical, electrocardiographic and electrophysiologic characteristics and long-term follow-up. J Am Coll Cardiol 1990; 15(6):1345-1354.

132. Shenasa H, Cooper RA, Pressley J et al. Site and origin of ectopic atrial tachycardia predicts its response to adenosine. J Am Coll Cardiol 1994; 23:250A(Abstract).

133. Griffith MJ, Linker NJ, Ward DE et al. Adenosine in the diagnosis of broad complex tachycardia. Lancet 1988; 1(8587):672-675.
134. Overholt ED, Rheuban KS, Gutgesell HP et al. Usefulness of adenosine for arrhythmias in infants and children. Am J Cardiol 1988; 61(4):336-340.
135. Wilber DJ, Baerman J, Olshansky B et al. Adenosine-sensitive ventricular tachycardia. Clinical characteristics and response to catheter ablation. Circulation 1993; 87(1):126-134.
136. Garratt CJ, Griffith MJ. Diagnosis of wide complex tachycardia. Pacing Clin Electrophysiol 1996; 19(5):878-879.
137. Rankin AC, Rae AP, Cobbe SM. Misuse of intravenous verapamil in patients with ventricular tachycardia. Lancet 1987; 2(8557):472-474.
138. Conti JB, Belardinelli L, Curtis AB. Usefulness of adenosine in diagnosis of tachyarrhythmias. Am J Cardiol 1995; 75(14):952-955.
139. Garratt CJ, Antoniou A, Griffith MJ et al. Use of intravenous adenosine in sinus rhythm as a diagnostic test for latent preexcitation. Am J Cardiol 1990; 65(13):868-873.
140. Ellenbogen KA, Rogers R, Damiano R. Utility of adenosine administration during intraoperative mapping in a patient with the Wolff-Parkinson-White syndrome. Pacing Clin Electrophysiol 1991; 14(6):985-988.
141. Drake I, Routledge PA, Richards R. Bronchospasm induced by intravenous adenosine. Hum Exp Toxicol 1994; 13(4):263-265.
142. Sylven C. Mechanisms of pain in angina pectoris—a critical review of the adenosine hypothesis. Cardiovasc Drugs Ther 1993; 7(5):745-759.
143. Sylven C, Crea F. Mechanism of anginal pain: the key role of adenosine. In: Belardinelli L, Pelleg A, eds. Adenosine and Adenine Nucleotides: From Molecular Biology to Integrative Physiology. Boston: Kluwer Academic Publishers, 1995:315-328.
144. Lerman BB, Wesley RC, Belardinelli L. Electrophysiologic effects of dipyridamole on atrioventricular nodal conduction and supraventricular tachycardia. Role of endogenous adenosine. Circulation 1989; 80(6):1536-1543.
145. Bertolet BD, Eagle DA, Conti JB et al. Bradycardia after heart transplantation: reversal with theophylline. J Am Coll Cardiol 1996; 28(2):396-399.
146. Ellenbogen KA, Thames MD, DiMarco JP et al. Electrophysiological effects of adenosine in the transplanted human heart. Evidence of supersensitivity. Circulation 1990; 81(3):821-828.
147. Haught WH, Bertolet BD, Conti JB et al. Theophylline reverses high-grade atrioventricular block resulting from cardiac transplant rejection. Am Heart J 1994; 128(6 Pt 1):1255-1257.

148. Saito D, Yamanari H, Matsubara K et al. Intravenous injection of adenosine triphosphate for assessing sinus node dysfunction in patients with sick sinus syndrome. Arzneimittelforschung 1993; 43(12):1313-1316.

149. Resh W, Feuer J, Wesley RC, Jr. Intravenous adenosine: a noninvasive diagnostic test for sick sinus syndrome. Pacing Clin Electrophysiol 1992; 15(11 Pt 2):2068-2073.

150. Alboni P, Ratto B, Cappato R et al. Clinical effects of oral theophylline in sick sinus syndrome. Am Heart J 1991; 122(5):1361-1367.

151. Raatikainen P, Dennis D, Guyton T et al. Differential rate-dependent effects of adenosine on AV nodal conduction and atrial repolarization: potential clinical implications. Circulation 1996; 94(8):I-284(Abstract).

152. Szentmiklosi AJ, Nemeth M, Szegi J et al. On the possible role of adenosine in the hypoxia-induced alterations of the electrical and mechanical activity of the atrial myocardium. Arch Int Pharmacodyn Ther 1979; 238(2):283-295.

153. Bertolet BD, Hill JA, Kerensky RA et al. Myocardial infarction-related atrial fibrillation: role of endogenous adenosine. Heart 1997; (in press).

154. Harrington GR, Froelich EG. Adenosine-induced torsades de pointes. Chest 1993; 103(4):1299-1301.

155. Wesley RC, Jr., Turnquest P. Torsades de pointe after intravenous adenosine in the presence of prolonged QT syndrome. Am Heart J 1992; 123(3):794-796.

156. Bertolet BD, Belardinelli L, Kerensky R et al. Adenosine blockade as primary therapy for ischemia-induced accelerated idioventricular rhythm: rationale and potential clinical application. Am Heart J 1994; 128(1):185-188.

157. Shah PK, Nalos P, Peter T. Atropine resistant post infarction complete AV block: possible role of adenosine and improvement with aminophylline. Am Heart J 1987; 113(1):194-195.

158. Lerman BB, Engelstein ED. Increased defibrillation threshold due to ventricular fibrillation duration. Potential mechanisms. J Electrocardiol 1995; 28 Suppl:21-24.

159. Wesley RC, Jr., Belardinelli L. Role of endogenous adenosine in postdefibrillation bradyarrhythmia and hemodynamic depression. Circulation 1989; 80(1):128-137.

160. Redmond JM, Zehr KJ, Gillinov MA et al. Use of theophylline for treatment of prolonged sinus node dysfunction in human orthotopic heart transplantation. J Heart Lung Transplant 1993; 12(1 Pt 1):133-8; discussion 138-9.

161. Lerman BB, Engelstein ED. Metabolic determinants of defibrillation. Role of adenosine. Circulation 1995; 91(3):838-844.

162. Viskin S, Belhassen B, Roth A et al. Aminophylline for bradyasystolic cardiac arrest refractory to atropine and epinephrine. Ann Intern Med 1993; 118(4):279-281.

163. Amoah-Apraku B, Xu J, Lu JY et al. Selective potentiation by an A_1 adenosine receptor enhancer of the negative dromotropic action of adenosine in the guinea pig heart. J Pharmacol Exp Ther 1993; 266(2):611-617.

164. Kollias-Baker C, Xu J, Pelleg A et al. Novel approach for enhancing atrioventricular nodal conduction delay mediated by endogenous adenosine. Circ Res 1994; 75(6):972-980.

165. Kollias-Baker C, Ruble J, Dennis D et al. Allosteric enhancer PD 81,723 acts by novel mechanism to potentiate cardiac actions of adenosine. Circ Res 1994; 75(6):961-971.

166. Mudumbi RV, Montamat SC, Bruns RF et al. Cardiac functional responses to adenosine by PD 81,723, an allosteric enhancer of the adenosine A_1 receptor. Am J Physiol 1993; 264(3 Pt 2):H1017-H1022.

167. Conti JB, Belardinelli L, Utterback DB et al. Endogenous adenosine is an antiarrhythmic agent. Circulation 1995; 91(6):1761-1767.

CHAPTER 4

Protective Effects of Adenosine in Reversibly and Irreversibly Injured Ischemic Myocardium

Robert D. Lasley

A denosine is a purine nucleoside which exerts numerous physiological effects in mammalian myocardium. These effects, which are described in detail in numerous excellent review articles,[1-3] include coronary vasodilation, negative chronotropic and dromotropic effects, and the antagonism of the metabolic and functional effects of β-adrenergic receptor stimulation. The majority of current evidence indicates that adenosine exerts no, or little, direct effects in ventricular myocardium.

In addition to exerting effects in normal myocardium, adenosine appears to exert numerous actions in ischemic and reperfused myocardium. Adenosine's effects in reperfused myocardium were recognized approximately 20 years ago when it was reported that the nucleoside could accelerate postischemic ATP resynthesis.[4,5] These observations, in fact, provided the initial rationale for the hypothesis that adenosine infusion could protect the ischemic heart. Only within the last 10 years, however, have the

Effects of Extracellular Adenosine and ATP on Cardiomyocytes,
edited by Amir Pelleg and Luiz Belardinelli. © 1998 R.G. Landes Company.

cardioprotective effects of adenosine been acknowledged. Adenosine infusion prior to ischemia delays the onset of ischemic contracture and reduces the rate of ATP catabolism. Adenosine also attenuates reversible postischemic left ventricular dysfunction, i.e., myocardial stunning, and reduces irreversible myocardial ischemic injury, i.e., infarct size. Similar to the actions of adenosine in normal myocardium, it appears that its cardioprotective effects are mediated via the stimulation of extracellular adenosine receptors located on specific cell types. Despite the acknowledged beneficial effects of adenosine in the ischemic heart, and the involvement of adenosine receptors, there is little definitive information on its mechanism of protection. This review will summarize the current knowledge of adenosine cardioprotection and address unresolved issues regarding potential mechanisms.

ADENOSINE METABOLISM AND MYOCARDIAL ISCHEMIA

This section will only cover the basic aspects of adenosine production in ischemic myocardium, which will provide the reader with the metabolic foundation for adenosine-mediated protection of the ischemic heart. Myocardial ischemia is defined as a reduction in coronary blood flow to the extent that the affected myocardium cannot maintain its normal contractile and metabolic state. With the onset of severe ischemia, ATP and creatine phosphate (CrP) catabolism exceed the rate of synthesis resulting in the net breakdown of these cellular energy metabolites. The catabolism of ATP results in increased AMP levels which is the source of the increased levels of the purine nucleoside adenosine observed in ischemic myocardium.

Adenosine may be formed from AMP by either cytosolic or membrane-localized 5'-nucleotidase. Evidence supporting ischemia-induced adenosine formation via cytosolic 5'-nucleotidase comes from studies in which intracellular adenosine is trapped as S-adenosyl homocysteine.[6-8] More recently Skladanowski et al[9] and Darvish and Metting[10] reported that the rates of adenosine production during ischemia could be accounted for by the activity and kinetic properties of cytosolic 5'-nucleotidase. Increased adenosine accumulation during ischemia could also arise from ecto 5'-nucleotidase. Extracellular AMP, which arises from the breakdown of ATP that is coreleased from nerve terminals with norepinephrine, can be metabolized to adenosine via ecto 5'-nucleotidase. Since catecholamines (and thus ATP) are released locally in ischemic myocardium this may be a source of increased adenosine production during ischemia. In fact Richardt et al[11] recently reported that adrenergic activation is responsible in part for adenosine production during myocardial ischemia.

Although the exact contributions of intracellularly- and extracellularly-derived adenosine to ischemia-induced increased adenosine production are not known, it is well known that adenosine is rapidly degraded to inosine by the enzyme adenosine deaminase. Inosine is then metabolized to hypoxanthine by nucleoside phosphorylase. In hearts in which xanthine oxidase is present, hypoxanthine is further metabolized to xanthine and uric acid. Figure 4.1 illustrates the relative increases in interstitial fluid (ISF) adenosine, inosine, and hypoxanthine in in vivo regionally ischemic canine myocardium in the absence (1a) and presence (1b) of the adenosine deaminase inhibitor erythro-9-(2-hydroxy-3-nonyl) adenine hydrochloride (EHNA).[12] Interstitial fluid metabolites were measured using the cardiac microdialysis technique,[13] which provides an estimate of metabolite concentrations in the extracellular compartment in the vicinity of the cardiac myocytes (i.e., ISF). Prior to coronary artery occlusion EHNA treatment increased dialysate adenosine concentration from 0.8 ± 0.1 to 2.5 ± 0.4 μM, and significantly decreased inosine and hypoxanthine levels. After 15 minutes ischemia dialysate adenosine levels in EHNA-treated animals were 30-fold greater than in control ischemic hearts. These results illustrate both the rapid metabolism of adenosine and the effectiveness of an adenosine metabolism inhibitor in intact myocardium.

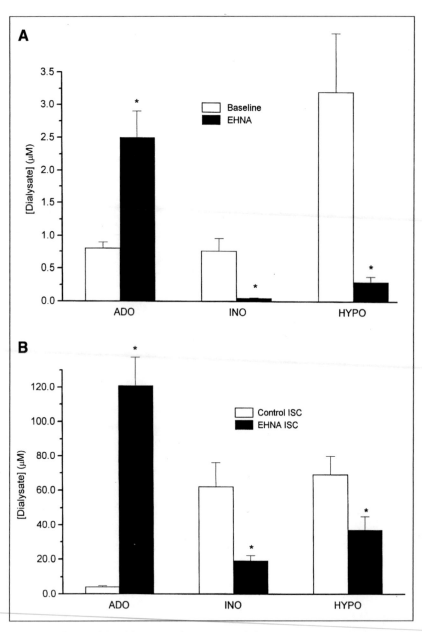

Fig. 4.1. Effects of the adenosine deaminase inhibitor EHNA on preischemic (A) and ischemic (ISC) (B) cardiac dialysate adenosine (ADO), inosine (INO) and hypoxanthine (HYPO) concentrations in the intact dog. Dialysate samples were collected with 1 cm exposed dialysis fibers perfused at 2 µl/min. Ischemia was induced by complete occlusion of the left anterior descending coronary artery. * $p < 0.05$ vs. baseline and control animals.

ADENOSINE AND POSTISCHEMIC
VENTRICULAR FUNCTION

The decrease in myocardial ATP, the rapid metabolism of adenosine, and the high levels of adenosine (and its metabolites) in the coronary venous outflow during reperfusion, led Benson et al[14] to suggest that the depressed levels of ATP during reperfusion were the result of a lack of precursors for adenine nucleotide resynthesis. The observations of Liu and Feinberg[15] and Wiedmeier et al,[16] that exogenous adenosine is rapidly incorporated into the myocardial adenine nucleotide pool, led to the hypothesis that treatment of the ischemic myocardium with exogenous adenosine could alleviate the metabolic and functional derangements that occurred after myocardial ischemia. Reibel and Rovetto[4] were the first investigators to directly test this hypothesis. Postischemic isolated perfused working rat hearts were reperfused for 30 minutes with 50 μM adenosine added to the perfusion medium. These investigators reported that adenosine had no beneficial effect on either left ventricular function or recovery of ATP levels. A subsequent study by the same authors[5] revealed that reperfusion for 5 hours with exogenous adenosine (50-200 μM) resulted in restoration of ATP to preischemic levels, but again they observed no improvement in the recovery of postischemic left ventricular function. These findings indicate that while adenosine can be incorporated into the adenine nucleotide pool in the postischemic heart (i.e., stimulation of purine salvage), this does not necessarily lead to improved postischemic cardiac function.

Subsequent studies on the cardioprotective effect of adenosine both by others[17,18] and in our laboratory[19,20] utilized a seemingly minor, but significant, modification in the treatment protocol. In these studies adenosine was administered prior to the onset of ischemia, rather than during reperfusion. We reported that isolated perfused rat hearts treated with adenosine (100 μM) prior to global normothermic ischemia exhibited preserved myocardial ATP contents and improved postischemic ventricular function.[19] The results of these studies combined with the initial studies by Reibel and Rovetto[4,5] indicated the importance of adenosine treatment prior to ischemia in promoting improved functional recovery.

Other investigators adopted a different approach to improving postischemic ventricular function with adenosine, specifically treating the heart with adenosine metabolism inhibitors. This approach was based on both the rapid metabolism of adenosine produced during ischemia and concerns about the systemic side-effects associated with high concentrations of exogenous adenosine. The first study utilizing this approach was performed by Foker et al[21] who treated dogs placed on cardiopulmonary bypass with the combination of adenosine and the adenosine deaminase inhibitor EHNA. Although this treatment regimen increased postischemic myocardial ATP levels, its effects on ventricular function were not assessed. Dhasmana et al[22] subsequently reported that isolated rat hearts treated with EHNA prior to ischemia exhibited improved postischemic ventricular function. Similar findings have been reported by others, in intact animals and isolated hearts.[12,23-25]

Consistent with the above findings that adenosine needed to be infused prior to ischemia to be protective, our initial studies in ischemic rat[19] and canine myocardium[20] indicated that adenosine pretreatment appeared to be exerting its cardioprotective effect during the period of ischemia. Isolated rat hearts treated with adenosine exhibited a 50% increase in the time to onset of ischemic contracture and a slower rate of ATP degradation during ischemia.[19] Dog hearts on cardiopulmonary bypass arrested with a crystalloid cardioplegic solution supplemented with adenosine also exhibited a slower rate of ATP breakdown during the period of arrest.[20] Since ATP resynthesis via purine salvage requires oxygen, this suggested to us that adenosine was exerting its metabolic effects, and probably its cardioprotective effect, via a mechanism other than ATP synthesis.

ADENOSINE RECEPTORS AND MYOCARDIAL ISCHEMIC INJURY

At the time of our initial observations of adenosine's effects in the ischemic heart, substantial evidence was accumulating that adenosine exerted its effects in various tissues by activating specific extracellular adenosine receptor subtypes. Initially, adenosine receptors were referred to as R_i and R_a depending on whether adenosine inhibited or activated, respectively, the synthesis of cAMP by adenylate cyclase.[26] Van Calker et al[27] then classified the two subtypes of adenosine receptors as A_1 and A_2 based on their relative affinities for structurally different adenosine analogs. By the early 1980s it was recognized that adenosine exerted its cardiac effects via the activation of receptors localized to different cell types.[28,29] In brief, adenosine-induced coronary vasodilatation is mediated via activation of adenosine A_2 receptors located on vascular smooth muscle and coronary endothelial cells. Adenosine's negative chronotropic, dromotropic and anti-adrenergic effects are mediated via the stimulation of adenosine A_1 receptors found primarily on cardiac myocytes. Since this original classification, myocardial A_2 receptors have been divided into at least two subtypes, and there is evidence for an A_3 receptor.

Work from our laboratory provided the first evidence that the cardioprotective effect of adenosine was mediated via the stimulation of a specific adenosine receptor subtype, the A_1 receptor.[30] Since we hypothesized that adenosine was exerting its beneficial effect during the period of ischemia, we tested the effects of different adenosine analogs on the time to onset of ischemic contracture (TOIC). Ischemic contracture is a rise in left ventricular resting pressure that occurs during severe ischemia, and is thought to reflect some aspect of ischemic injury. Paced, constant flow perfused isolated rat hearts were instrumented with a latex balloon in the left ventricle and subjected to global, zero flow normothermic ischemia. Control hearts were compared to those treated prior to ischemia with adenosine (100 μM), the adenosine A_1 receptor agonist N6-(2-phenylisopropyl)-adenosine (PIA, 1 μM), the A_2 agonist phenylaminoadenosine (PAA, 1 μM), and adenosine or PIA in the presence of the adenosine receptor antagonist BW A1433U, a para-phenyl carboxyl-substituted derivative of 1,3-dipropyl-8-phenylxanthine (10 μM). Hearts treated with adenosine and PIA exhibited TOICs of 18.6 ± 0.4 and 16.6 ± 1.2 minutes, respectively, compared to control hearts (9.1 ± 0.7 min). In contrast, treatment with PAA had no significant effect (TOIC = 11.7 ± 0.9 min). The beneficial effects of adenosine and PIA were blocked by the adenosine receptor blocker BW A1433U. Treatment with BW A1433U alone shortened TOIC to one-half that of control hearts (4.7 ± 0.5 min, $p < 0.01$). These results indicated that exogenous and endogenous adenosine protected the ischemic heart via an adenosine A_1 receptor mechanism. Since A_1 receptors are located predominantly on the cardiac myocytes, these results were consistent with the hypothesis that adenosine was protecting the ischemic myocyte.

ADENOSINE RECEPTORS AND POSTISCHEMIC VENTRICULAR FUNCTION

Subsequent studies from several laboratories[31-33] indicated that infusion of adenosine A_1 receptor agonists prior to ischemia also improved postischemic recovery of function. In our isolated rat heart preparation of global ischemia,[31] adenosine cardioprotection is mimicked by the adenosine A_1 receptor analog cyclohexyladenosine (CHA, 0.25 µM), but not by the adenosine A_2 receptor agonist PAA (0.25 µM) (Fig. 4.2). Activation of A_1 receptors by CHA was confirmed by a reduction in spontaneous heart rate (from 275 ± 9 to 76 ± 12 beats/min), whereas PAA had no effect on heart rate. As in all of our experiments with adenosine A_1 receptor agonists, hearts were paced to exclude adenosine's negative chronotropic effect as a cardioprotective mechanism. The protective effect of adenosine was blocked by the selective adenosine A_1 receptor blocker 8-cyclopentyl-1,3-dipropylxanthine (DPCPX, 5 µM). The selectivity of DPCPX for the A_1 receptor was evident during the adenosine + DPCPX treatment period as spontaneous heart rate remained unchanged (266 ± 4 vs. 277 ± 4 beats/min), but coronary flow increased 64%.

Although there is substantial evidence in isolated hearts that adenosine attenuates postischemic ventricular dysfunction via an adenosine A_1 receptor mechanism, there is only one study to date that directly supports this hypothesis in in vivo myocardium. Yao and Gross[32] reported that the adenosine A_1 receptor agonist cyclopentyladenosine (CPA), but not the A_2 receptor agonist CGS 21680 (2-[p-(2-carboxyethyl)-phenethylamino]-5'-N-ethylcar-

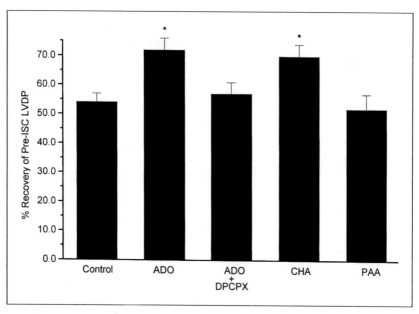

Fig. 4.2. Effects of adenosine (ADO) + the adenosine A_1 receptor antagonist DPCPX and ADO receptor agonists on postischemic function in isolated perfused rat hearts. Hearts were submitted to 30 minutes global normothermic ischemia (ISC) and 45 minutes reperfusion. Results are expressed as % recovery of pre-ISC left ventricular developed pressure (LVDP). All agents were administered for 10 minutes immediately prior to the onset of ISC. Abbreviations are as described in the text. * $p < 0.05$ vs. control hearts.

boxamidoadenosine), attenuated regional myocardial stunning (induced by brief repetitive coronary occlusions) in an in vivo canine preparation. The beneficial effect of CPA (2.0 µg/kg/min, iv infusion for 15 minutes) was not due to any effects on systemic hemodynamics, reduction in heart rate, increase in coronary blood flow, or increase in coronary collateral blood flow during ischemia. Although the combination of CPA and the selective adenosine A_1 receptor antagonist DPCPX was not tested, DPCPX alone actually exacerbated myocardial stunning. Another interesting observation in this study was that both CPA and DPCPX had to be administered prior to the first coronary occlusion to exert their respective beneficial and deleterious effects. These results are consistent with the hypothesis that adenosine exerts its cardioprotective effect during the early period of myocardial ischemia via an adenosine A_1 receptor mechanism. However, there remains the need for more experimental studies in in vivo preparations in other species to determine the exact role of adenosine receptors in attenuating myocardial stunning.

Although adenosine is cardioprotective, it has been recognized for several years that relatively high doses of adenosine must be infused in order to exert a protective effect. Concentrations of approximately 10 μM adenosine in isolated hearts and 50 μg/kg/min (intracoronary) in intact preparations appear to be the minimum effective doses of adenosine to attenuate myocardial stunning. These doses of adenosine are considerably greater than those needed to induce maximal coronary vasodilation in both types of preparations. This is most likely due to the rapid metabolism of adenosine by red blood cells and endothelial cells which produces a large concentration gradient between intravascular and interstitial fluid (ISF) adenosine concentrations. Since the ISF bathes the myocytes, adenosine concentrations in this compartment may be the most accurate determinant of the extent to which the myocyte A_1 receptor is activated. It follows then that in order for adenosine to attenuate myocardial stunning, ISF adenosine levels must be elevated above basal levels prior to or during ischemia.

Using a modification of the brain microdialysis technique to measure cardiac ISF adenosine and its metabolites,[13] we have obtained results from isolated heart and in situ preparations consistent with the above hypothesis. The dialysate fluid samples collected from the outflow of the dialysis fiber provide an index of ISF metabolites. In an in situ pig preparation intracoronary adenosine infused at a concentration of 5.0 μg/kg/min increased coronary venous plasma adenosine levels from 0.31 ± 0.09 to 1.51 ± 0.32 μM. Dialysate adenosine levels increased only slightly (from 0.55 ± 0.05 μM to 0.71 ± 0.11 μM). Increasing the adenosine infusion rate to 50 μg/kg/min resulted in further increases in venous (15.16 ± 1.80 μM) and dialysate (2.02 ± 0.09 μM) adenosine concentrations. These findings demonstrate the tremendous metabolic barrier to exogenous adenosine and the importance of measuring adenosine levels in the appropriate extracellular fluid compartment. In in vivo canine[34] and porcine[35] experiments in our laboratory a dose of 50 μg/kg/min infused immediately prior to coronary occlusion increases ISF adenosine levels and attenuates regional myocardial stunning by 50%. Similarly, preischemic administration of the adenosine deaminase inhibitors EHNA[12] and pentostatin[25] (deoxycoformycin) in dogs is associated with elevated preischemic ISF adenosine (measured by cardiac microdialysis) and improved regional ventricular function.

Additional studies in our laboratory further support the links between increased preischemic ISF adenosine, myocyte A_1 receptor stimulation and improved postischemic ventricular function. Figure 4.3 illustrates the effects of increasing concentrations of exogenous adenosine on spontaneous heart rate and dialysate adenosine in the isolated perfused rat heart. Infusion of 1 μM adenosine had no effect on heart rate or dialysate adenosine, whereas 10 and 100 μM adenosine decreased heart rate and dose-dependently increased dialysate adenosine concentrations. All three doses of adenosine produced increases in coronary effluent adenosine levels and maximal coronary dilation, consistent with A_2 receptor activation. The two doses that increased dialysate adenosine reduced spontaneous heart rate, i.e., A_1 receptor activation. Consistent with the hypothesis of adenosine A_1 receptor-mediated cardioprotection, 10 and 100 μM adenosine, but not 1 μM adenosine, significantly improved postischemic ventricular function.

As originally observed by Reibel and Rovetto[4,5] it appears that adenosine must be administered prior to ischemia to attenuate postischemic ventricular dysfunction. More recent reports in in vivo preparations indicate that adenosine infusions during reperfusion do not attenuate myocardial stunning.[34,36] We[34] observed a transient increase in function accompanying intracoronary adenosine infusion during reperfusion in dogs, but this effect rapidly dissipated following termination of the infusion with the same time course that adenosine-induced hyperemia subsided. It also appears that the pretreatment adenosine infusion must be maintained until the onset of ischemia in order to improve postischemic function.

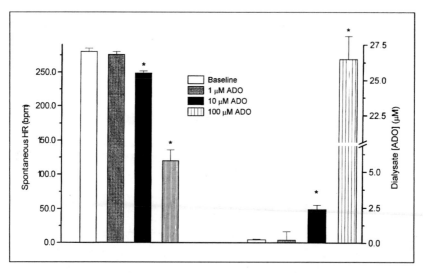

Fig. 4.3. Exogenous adenosine (ADO) effects on spontaneous heart rate (HR) and dialysate ADO concentration ([ADO]) in isolated perfused rat hearts. Separate groups of hearts were treated with either 1, 10 or 100 DM for 10 minutes. Dialysate samples were collected with 0.8 cm exposed dialysis fibers perfused at 0.75 µl/min for 10 minutes. * $p < 0.05$ vs. control hearts.

There are reports in four species (rabbit, dog, pig and rat) that a transient adenosine infusion (adenosine preconditioning) that is terminated prior to ischemia does not enhance postischemic function.[33,36-38]

In summary, there is significant experimental evidence that adenosine improves postischemic ventricular function. This cardioprotective effect is dependent on the administration of adenosine prior to ischemia in a concentration sufficient to reach the myocardial ISF which bathes the cardiac myocytes. Elevation of endogenous adenosine with adenosine deaminase inhibitors also attenuates myocardial stunning. Finally, results obtained with adenosine receptor agonists and the adenosine A_1 receptor antagonist DPCPX indicate that adenosine attenuation of myocardial stunning appears to be mediated via activation of cardiac myocyte adenosine A_1 receptors.

ADENOSINE AND MYOCARDIAL INFARCT SIZE

Reperfusion Adenosine Treatment

Although the initial studies on the cardioprotective effects of adenosine focused on its ability to enhance postischemic function, substantial evidence has accumulated that adenosine also reduces infarct size. The initial studies on adenosine's effects in the irreversibly injured heart tested the effects of adenosine infusions during reperfusion. Olafsson et al[39] infused intracoronary adenosine at a dose of approximately 150 μg/kg/min for the first hour of reperfusion after a 90-minute coronary occlusion in intact dogs. They reported that adenosine treatment was associated with a 75% reduction in infarct size after 24 hours reperfusion. A subsequent study by the same laboratory[40] showed that the same dose of adenosine administered intravenously for the first 2.5 hours of reperfusion following a 90-minute occlusion also reduced myocardial infarct size in the dog after 72 hours reperfusion. In both of these studies the ischemic beds treated with adenosine exhibited significantly less neutrophil accumulation and erythrocyte plugging of capillaries. These latter two observations are consistent with adenosine's ability to inhibit both neutrophil adherence to endothelium[41] and platelet aggregation.[42]

In contrast to the above reports, there are additional studies which indicate that adenosine infusion during reperfusion does not reduce infarct size. Homeister et al[43] observed that the results of Olafsson et al[39] were obtained in animals treated with lidocaine. When these studies were repeated (also in dogs) adenosine and lidocaine in combination, but neither agent alone, reduced infarct size. More recently Vander Heide and Reimer[44] reported that, even in the presence of lidocaine, adenosine administered during reperfusion did not reduce infarct size in the dog. Similar discrepant results have been reported in the rabbit. Norton et al[45] reported that three different doses of adenosine infused intravenously during reperfusion could reduce infarct size. However, Goto et al[46] reported that, even in the presence of lidocaine, reperfusion adenosine (150 and 370 µg/kg/min) did not reduce myocardial infarct size in the rabbit. It is not clear why these differences exist since these conflicting reports in both dogs and rabbits have been obtained with the same doses of adenosine.

Further clouding the above controversy are the results obtained with adenosine receptor analogs infused during coronary reperfusion. Norton et al[47] reported that both the adenosine A_1 receptor analog cyclopentyladenosine (CPA) and the adenosine A_2 receptor agonist CGS 21680 infused during reperfusion significantly reduced infarct size (measured after 48 hours reperfusion) in the rabbit. In contrast, Thornton et al[48] reported that in their rabbit model, reperfusion treatment with the adenosine A_1 agonist PIA did not reduce infarct size. However, Norton et al[47] used low doses (0.01-10 µg/min) infused for the first hour of reperfusion, whereas Thornton et al[48] used a 5-minute infusion of a much higher dose (approx. 360 µg/min). The latter regimen was associated with profound systemic hypotension (25 mm Hg) after the first hour of reperfusion, whereas only the highest CPA dose used in the former study had any effect on blood pressure (a modest decrease). There have been few additional studies of the effects of reperfusion treatments with adenosine A_1 receptor agonists on myocardial infarct size, but there are other reports of the beneficial effects of adenosine A_2 agonists administered during reperfusion.[49,50] Jordan et al[50] recently reported that the infusion of a low dose (2 µg/kg/min) of the A_2 analog CGS 21680 reduced infarct size in the dog by nearly 50%. The reduction in infarct size was associated with decreased neutrophil accumulation, and, in in vitro studies, CGS 21680 reduced neutrophil adherence to endothelium and superoxide anion production by activated neutrophils.

The reports of the beneficial effects of reperfusion adenosine and adenosine A_2 receptor agonists are consistent the anti-inflammatory effects of adenosine A_2 receptor activation in polymorphonuclear leukocytes (PMN).[41,51] However it has also been reported that PMN adenosine A_1 receptor activation is associated with the opposite effects, i.e., promotion of neutrophil adherence and superoxide generation.[41,51] In fact there are at least two reports that these latter effects may contribute to myocardial reperfusion injury.[52,53] Thus it is possible that the opposing effects of PMN adenosine A_1 and A_2 receptor activation may explain some of the confusion regarding the effects of adenosine and adenosine receptor agonists on infarct size when administered during reperfusion.

PRESCHEMIC ADENOSINE TREATMENT AND INFARCT SIZE

Although the question of whether adenosine administered during coronary reperfusion remains unresolved, there is substantial evidence that adenosine infused prior to ischemia reduces infarct size. Preischemic intracoronary and intravenous adenosine and adenosine A_1 receptor agonist infusions have been shown to reduce infarct size in dogs,[54] pigs[55] and rabbits.[48,56,57] Interestingly, though there are numerous reports that adenosine improves postischemic function in rat myocardium, there are few, if any, reports of adenosine reducing infarct size in this species. In fact two studies reported that neither adenosine deaminase inhibition, adenosine, nor the adenosine A_1 receptor agonist PIA reduced infarct size in the rat.[58,59] Liu and Downey[60] reported that the infusion of the adenosine A_1 receptor agonist CCPA reduced infarct size by 50% in the isolated rat heart, but adenosine, similar to the results of Li and Kloner[58] these authors reported severe hypotension with PIA and in the intact rat. Similar to its anti-stunning effect, adenosine appears to reduce myocardial infarct size by an adenosine A_1 receptor mechanism, since the preischemic infusion of adenosine A_2 receptor agonists does not reduce infarct size,[48] and the beneficial effects of adenosine and/or its A_1 agonists are blocked by the relatively selective adenosine A_1 receptor antagonist DPCPX.[57,61]

ENDOGENOUS ADENOSINE AND MYOCARDIAL INFARCT SIZE

In addition to the above reports with exogenous adenosine, there is evidence that endogenous adenosine may modulate myocardial infarct size. In fact, the majority of the studies on adenosine's beneficial effects in the irreversibly injured heart were based on initial reports implicating the cardioprotective effects of endogenous adenosine. Liu et al[62] reported that the adenosine receptor antagonists SPT and PD 115,199 blocked the infarct size reducing effect of ischemic preconditioning. Ischemic preconditioning is the phenomenon whereby a brief period of myocardial ischemia and reperfusion reduces infarct size following a subsequent prolonged occlusion.[63] This hypothesis is further supported by observations that the brief preconditioning ischemia is associated with increased ISF adenosine concentration.[57,64] Brief infusions of adenosine and adenosine A_1 agonists, which are terminated prior to the long coronary occlusion, also mimic ischemic preconditioning.[33,48,54,56-60,62]

There are conflicting reports however on the infarct size reducing effect of pharmacological agents which increase endogenous adenosine levels. It has been reported that treatments with the nucleoside transport blockers dipyridamole[65] and R75231 (draflazine)[66] and the adenosine deaminase inhibitor pentostatin[67] do not reduce infarct size. These same agents however have been shown to potentiate the effects of ischemic preconditioning.[65-67] It is possible that these agents alone did not increase ISF adenosine prior to ischemia in nonpreconditioned hearts, but did so in combination with the brief preconditioning ischemia. In all but one of the above studies however, neither plasma nor ISF adenosine levels were measured. Silva et al[67] reported that pentostatin treatment in nonpreconditioned dogs was associated with increased ISF adenosine prior to and during coronary occlusion, but infarct size was not reduced. However, this lack of protection could have been due to coronary steal, since reactive hyperemia was reduced and coronary flow in the nonischemic bed was significantly increased in pentostatin-treated animals.

Results obtained in our laboratory in the in situ pig indicate that the nucleoside transport blocker R75231, when administered prior to ischemia, can reduce myocardial infarct size.[68] The infusion of R75231 (0.1 mg/kg, i.v.) increased preischemic ISF adenosine concentration from 0.34 ± 0.04 to 0.73 ± 0.08 μM. During the 60-minute coronary occlusion ISF adenosine was significantly increased and inosine and hypoxanthine levels were reduced. As illustrated in Figure 4.4, after 2 hours reperfusion infarct size in the R75231-treated pigs was reduced from $38.4 \pm 2.6\%$ of the area at

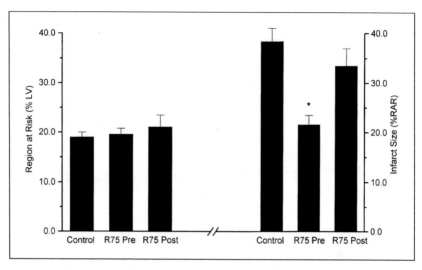

Fig. 4.4. Effect of preischemic (Pre) and postischemic (Post) treatments with the nucleoside transport inhibitor R75231 on infarct size in an in vivo pig preparation. Treatments (0.1 mg/kg) were administered 15 minutes prior (Pre) to coronary artery occlusion or 15 minutes prior to reperfusion (Post). Infarct size was estimated after two hours reperfusion with triphenyl tetrazolium chloride (TTC) staining. The region at risk (RAR) is expressed as percent of the left ventricle (% LV) and infarct size as percent RAR. * p < 0.05 vs. control hearts.

risk in control animals to 21.6 ± 1.9%. When the same dose of R75231 was administered 15 minutes prior to reperfusion, both ISF and coronary venous plasma adenosine levels were elevated during reflow. In reperfusion R75231 treated pigs, coronary venous plasma adenosine concentration was 2-3 times greater than that in control pigs for the first 30 minutes of reperfusion, but infarct size was not altered (33.5 ± 3.5% of risk area). These results are consistent with the hypothesis that a preischemic elevation of ISF adenosine concentration reduces infarct size, but that elevating endogenous adenosine levels during reperfusion is not beneficial.

Our findings with R75231 reperfusion treatment are consistent with the reports that adenosine infusion during reperfusion does not reduce infarct size.[43,44,46] However they are at odds with another report that modulation of endogenous adenosine during reperfusion may reduce infarct size. Zhao et al[69] reported that the reperfusion administrations of the adenosine receptor antagonists

8-SPT and PD 115,199 increased myocardial infarct size in the rabbit. These authors also reported that if 8-SPT was infused after the first 30 minutes of reperfusion infarct size was not altered.[69] This latter observation is consistent with our findings in the intact pig that coronary venous adenosine concentration in control and R75231-treated animals returned to basal levels between 30 and 60 minutes of reperfusion.[68] One explanation for these seemingly contradictory findings is that if endogenous adenosine is itself cardioprotective, then elevating adenosine levels further may not reduce infarct size any further.

Although the consistent infarct size reducing effect of adenosine pretreatment has received the most attention, the discrepant reports of adenosine's effects during reperfusion warrant more study. Since endogenous vascular adenosine appears to modulate infarct size for the first 30 minutes of reperfusion (based on the results of Zhao et al[69]), in order to decrease infarct size any further with adenosine, it may be necessary to maintain elevated levels for a longer duration of reperfusion. The lack of plasma adenosine measurements in these studies and the use of different doses of adenosine receptor blockers may explain some of these discrepancies. However, there are enough studies documenting the anti-inflammatory effects of adenosine A_2 receptor activation to suggest that manipulation of adenosine levels and receptors during reperfusion may modulate myocardial infarct size. Another reason for addressing this issue further is that the reperfusion administration of adenosine alone or in combination with an adenosine potentiator, is more clinically relevant in the setting of acute myocardial infarction, than is adenosine pretreatment.

ADENOSINE AND ISCHEMIC PRECONDITIONING

As mentioned above, the majority of the studies on endogenous adenosine's modulation of myocardial infarct size have been based on reports implicating adenosine's role in the cardioprotective effect of ischemic preconditioning. Although ISF adenosine levels do increase during the brief preconditioning occlusion,[57,64] and adenosine can mimic the infarct size reducing effect of ischemic preconditioning,[48,54-57] there are differences between the cardio-protective effects of adenosine and ischemic preconditioning. One of these differences, although only initially appearing to be a minor variation in protocol, has quite disparate outcomes. Adenosine pretreatment, in which the adenosine infusion is not terminated until the onset of ischemia, reduces both myocardial stunning and infarct size.[31-36] In contrast, a transient adenosine infusion that is terminated prior to ischemia (adenosine preconditioning) reduces infarct size,[48,54-57] but does not attenuate postischemic dysfunction.[33,36] Ischemic preconditioning, but not adenosine preconditioning, reduces the rate of adenine nucleotide catabolism during the prolonged ischemia.[57,64]

The most striking difference between the beneficial effects of adenosine and ischemic preconditioning on infarct size is their susceptibility to blockade by the selective adenosine receptor antagonist DPCPX. This discrepancy in rabbit myocardium is illustrated in Figure 4.5. Isolated rabbit hearts were submitted to 45 minutes global normothermic ischemia and one hour reperfusion.[33] Control hearts were compared to hearts pretreated with the adenosine A_1 agonist PIA (1 µM) or preconditioned with 5 minutes ischemia/10 minutes reperfusion. The two treatments were then repeated in the presence of DPCPX (2.5 µM). The infarct size reducing effect of PIA in this preparation was blocked by DPCPX, but that of ischemic preconditioning was not altered. A similar inability of DPCPX to block ischemic preconditioning in the regionally ischemic isolated rabbit heart has been reported by Liu and Downey.[70] Although DPCPX has been reported to block ischemic preconditioning in the dog,[54] this antagonist also does not block hypoxic preconditioning in the rat.[71]

In summary, there is significant experimental evidence that adenosine attenuates myocardial stunning and reduces infarct size when administered prior to ischemia, presumably by myocyte adenosine A_1 receptor activation. When infused during reperfusion, adenosine does not produce a sustained improvement in function, and although there are discrepant results on its ability to reduce infarct size under these conditions, there is both in vitro and in

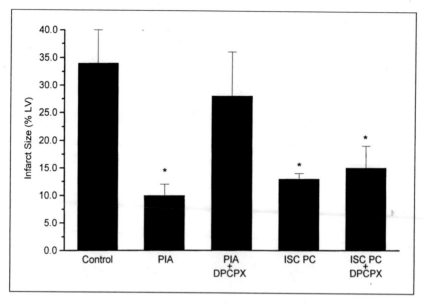

Fig. 4.5. Comparison of adenosine A_1 receptor antagonist, DPCPX, effects on infarct size reduction with the adenosine A_1 receptor agonist PIA and ischemic preconditioning (ISC PC) in the isolated perfused rabbit heart. Hearts were submitted to 45 minutes global normothermic ischemia (ISC) and 60 minutes reperfusion. Infarct size, expressed as % of the left ventricle (LV), was then estimated with TTC staining. Treatment with PIA ± DPCPX was limited to the 5 minutes immediately preceding ISC. Ischemic PC was induced with 5 minutes global ISC and 10 minutes reperfusion prior to 45 minutes ISC. In the ISC PC + DPCPX group DPCPX was administered 10 minutes prior to the 5 minutes ISC. * $p < 0.05$ vs. control hearts.

vivo evidence that adenosine A_2 receptor stimulation may reduce myocardial infarct size. Adenosine also appears to play some role in ischemic preconditioning, although to what extent and what receptors are involved remain to be determined.

MECHANISMS OF ADENOSINE-MEDIATED CARDIOPROTECTION

Although there is significant evidence indicating that adenosine is cardioprotective, there is much less known about its mechanism of action. This is due, in part, to the fact that in normal ventricular myocytes adenosine exerts no significant direct effects on metabolism, intracellular calcium, or contractility. Even under simulated ischemia conditions in cardiac myocytes, adenosine does not exert the same extent of protection as it does in the whole heart.[71] These observations suggest that adenosine cardioprotection may be mediated via adenosine receptor modulation of responses to other autocrine or paracrine factors that are released during ischemia and/or reperfusion.

Adenosine, when administered prior to ischemia, reduces stunning and infarct size via adenosine A_1 receptor activation. Although one would presume that this protection occurs via similar mechanisms, this may not be the case since, as described above, adenosine preconditioning reduces infarct size, but does not attenuate stunning. Whether this is due to different mechanisms of protection or merely due to differences in the development of reversible vs. irreversible injury remains to be determined. Although initial theories regarding adenosine cardioprotection were based on increasing adenosine levels during the ischemic period, the results of more recent studies, in which ISF adenosine levels were reported, suggest that it is the preischemic increase in adenosine, and presumably the preischemic activation of the A_1 receptor, that is the key factor. Studies in our laboratory on the cardioprotective effects of exogenous and endogenous adenosine in preparations of stunning and infarction are associated with a preischemic increase in ISF adenosine,[12,19,31,34,35,57] but adenosine levels during ischemia in treated animals are not consistently elevated. The exact significance of this observation is not known, but it does stress the importance of elevating adenosine levels prior to ischemia.

Although there is evidence that adenosine, when administered during reperfusion, may modulate myocardial infarct size via adenosine A_2 receptor activation, there is little definitive information on this mechanism. This most likely is due to the controversial results obtained with this treatment protocol. In contrast there have been many studies addressing the mechanism of adenosine A_1 receptor mediated cardioprotection. The section below will address this aspect of adenosine cardioprotection.

PROPOSED MECHANISM(S) OF ADENOSINE A_1 RECEPTOR MEDIATED CARDIOPROTECTION

It has been well established that adenosine's effects on the time-independent potassium current (I_{KADO}) in atrial cells and its anti-adrenergic effect in ventricular myocardium are mediated by A_1 receptor coupling to an inhibitory guanine nucleotide binding protein (presumably G_i). This hypothesis is supported by the results of studies in which pertussis toxin pretreatment, which catalyzes the ADP ribosylation of the α subunit of G_i, thus rendering it inactive, attenuates the negative chronotropic and anti-adrenergic effects of adenosine (see ref. 73 for a review). It also appears that adenosine's beneficial effect on postischemic ventricular function may occur via a similar mechanism. We[74] have reported that in rats pretreated with pertussis toxin (25 µg/kg i.p., 48 hours prior to isolation) neither adenosine nor CHA infusions enhanced postischemic function in the isolated hearts after global ischemia (Fig. 4.6). We did not measure ADP ribosylation, but tested the effective inactivation of G_i by measuring the effects of adenosine and CHA on spontaneous heart rate. In normal hearts adenosine (100 µM) and CHA (0.25 µM) reduced spontaneous heart rates from 275 ± 7 to 100 ± 10 and 275 ± 9 to 76 ± 12 beats/min, respectively. However, in pertussis toxin pretreated animals adenosine (286 ± 9 to 251 ± 15 beats/min) and CHA (268 ± 7 to 232 ± 10 beats/min) had only minor effects on spontaneous heart rate, suggesting that pertussis toxin sensitive G_i proteins were disabled. We also observed that the preischemic infusion of the muscarinic receptor agonist carbachol, which produces many of the same cardiac actions as adenosine A_1 receptor agonists, improved postischemic function similar to adenosine and CHA (Fig. 4.6). These results suggest that adenosine pretreatment improves postischemic ventricular function via A_1 receptor coupling to pertussis toxin-sensitive G_i proteins.

Since adenosine and its A_1 receptor analogs both attenuate stunning and reduce infarct size, it is possible that adenosine's anti-infarct also occurs via a pertussis toxin sensitive G protein. However, there are no studies directly testing this hypothesis. Thornton

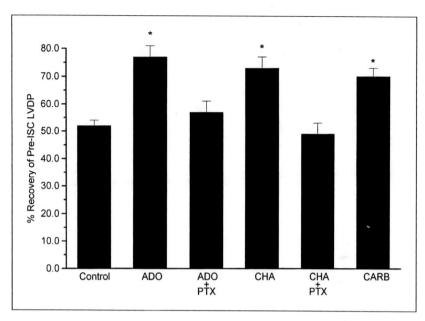

Fig. 4.6. Pertussis toxin (PTX) blockade of the cardioprotective effects of adenosine (ADO) and the ADO A_1 receptor agonist CHA in isolated perfused rat hearts. The protocol, treatment regimens and expression of results are the same as described in Fig. 4.2. In the PTX groups rats were treated with PTX 48 hours prior to the isolated heart experiments. Carbachol (CARB) was infused 10 minutes prior to ischemia (ISC). * $p < 0.05$ vs. control hearts.

et al[75] reported that the ability of carbachol to reduce infarct size in the regionally ischemic isolated rabbit heart preparation was abolished in animals pretreated with pertussis toxin. Ischemic preconditioning, which appears to be mediated in part by adenosine, was also blocked in pertussis toxin treated rabbits.[75] Since carbachol and adenosine exert similar effects in ventricular myocardium, and adenosine mimics the infarct size reducing effect of ischemic preconditioning, it is likely that adenosine also reduces infarct size via A_1 receptor coupling to pertussis toxin sensitive G_i proteins.

With respect to mechanism(s) of adenosine cardioprotection distal to G protein coupling, there is even less definitive information. One proposed mechanism that is supported by several lines of evidence is adenosine's anti-adrenergic effect. Adenosine and adenosine A_1 receptor agonists inhibit catecholamine release[76] and attenuate the functional and metabolic effects of β-adrenergic stimulation in the nonischemic heart.[77] There is also evidence that not only do exogenous and endogenous adenosine reduce nonexocytotic norepinephrine release during myocardial ischemia,[78] but that myocardial catecholamine release contributes to ischemia-induced increases in adenosine.[11] Consistent with these reports are two recent studies in intact[79] and isolated hearts,[80] whose authors concluded that adenosine was cardioprotective by exerting an anti-adrenergic effect during the period of ischemia. These findings suggest that adenosine's anti-adrenergic properties may play a role in its cardioprotective effects.

ADENOSINE CARDIOPROTECTION AND ATP-SENSITIVE K+ CHANNELS

Another signal transduction pathway implicated in adenosine cardioprotection is the activation of ATP-sensitive potassium (K_{ATP}) channels. Support for this mechanism is provided by observations that adenosine reductions of postischemic dysfunction[32] and infarct size[54] in the dog and infarct size in the pig[55] can be abolished by the K_{ATP} channel blocker glibenclamide and mimicked by K_{ATP} channel openers. Glibenclamide also blocks the infarct size-reducing effect of adenosine in the rabbit.[81] Yao et al[82] reported a synergistic interaction between adenosine and the potassium channel opener bimikalim in a canine infarct model of adenosine preconditioning. There are also reports that adenosine and A_1 receptor agonists activate or potentiate K_{ATP} channel currents in isolated ventricular myocytes.[83-85]

Although there is compelling evidence coupling adenosine A_1 receptor mediated cardioprotection to K_{ATP} channel activation, there are several aspects of this theory which must be addressed. First, the specific link between adenosine A_1 receptors and the K_{ATP} channel has yet to be determined. In a frequently cited study, Kirsch et al[83] reported that adenosine A_1 receptors were coupled to K_{ATP} channels in 1-3 day old neonatal rat ventricular myocytes. However these results were obtained in only 7 of 16 experiments under low ATP conditions, and Babenko and Vassort[86] recently reported a lack of effect of adenosine on K_{ATP} channel current in adult rat ventricular myocytes. This latter observation is consistent with the report that the K_{ATP} channel blocker glibenclamide does not block adenosine-mediated attenuation of postischemic dysfunction in the isolated rat heart.[87] Liu et al[85] reported that in rabbit ventricular myocytes exogenous adenosine did not directly activate K_{ATP} even under simulated ischemia conditions, but glibenclamide has been reported to block adenosine-induced reduction of myocardial infarct size in this same species.[81]

These discrepant results concerning the link between adenosine and K_{ATP} channels in ventricular myocardium could be due to numerous effects, in addition to species differences. One possibility is that adenosine does not directly activate this channel, but modulates its sensitivity to other endogenous or exogenous activators during ischemia. This may explain the results of studies in adult isolated ventricular myocytes in which adenosine A_1 agonists did not activate I_{KATP}.[85,86] The lack of a direct effect of adenosine on K_{ATP} channels is also consistent with the fact that adenosine and adenosine A_1 agonists do not produce the same effects in ventricular myocardium as do K_{ATP} channel openers.[88] Differences between the effects of adenosine A_1 receptor stimulation and K_{ATP} channel openers are not unexpected, since K_{ATP} channel openers bind directly to the channel, but there is no evidence that adenosine does so.

Another factor which must be recognized is that the majority of the evidence implicating K_{ATP} channel activation in adenosine A_1 receptor-mediated cardioprotection is based on results obtained with the K_{ATP} channel blocker glibenclamide. However, glibenclamide blocks the cardiac effects of K_{ATP} channel openers at doses significantly lower than those required to block the effects of adenosine.[89] Glibenclamide also blocks adenosine-induced coronary dilation at doses much lower than those used to block adenosine cardioprotection.[90] It has been reported that glibenclamide, at concentrations greater than 1 μM, blocks the effects of adenosine on I_{KADO} in guinea pig atrial myocytes.[89] Thus K_{ATP} channel-independent effects of glibenclamide must be recognized when linking adenosine A_1 receptors and I_{KATP}.

ADENOSINE CARDIOPROTECTION
AND PROTEIN KINASE C

Another proposed signal transduction mechanism for adenosine mediated cardioprotection is the activation of calcium dependent protein kinase (PKC). This hypothesis is based on observations that ischemic preconditioning in rabbits is blocked by the adenosine receptor antagonist SPT and PKC inhibitors,[62,91,92] and ischemic preconditioning can be mimicked by adenosine A_1 receptor agonists.[48,62] However, there are no studies to date showing the effects of adenosine or A_1 receptor agonists on PKC isoform activity in ischemic ventricular myocardium. In fact, Armstrong et al[93] reported that adenosine preconditioning in rabbit isolated ventricular myocytes did not induce translocation of any PKC isoform studied. Translocation of PKC from the cytosol to the plasma membrane is considered an index of PKC activation. In light of these observations, the significance of the report by Sakamoto et al,[92] that the infarct size reducing effect of PIA in the intact rabbit could be blocked by the PKC inhibitor staurosporine, remains to be determined.

Henry et al[94] reported that treatments of normoxic rat ventricular myocytes with adenosine and the adenosine A_1 receptor analog PIA were associated with a transient cytosol to membrane translocation of the δ-PKC isoform, and both effects were blocked by DPCPX. The physiological significance of this observation, however, is not clear, since PKC translocation peaked at 1 minute and then decreased to basal levels within 5-10 minutes, despite the continued presence of adenosine or PIA. In addition adenosine A_1 receptor activation in ventricular myocardium is not associated with any effects typically seen with receptor agonists that couple to phospholipase C (PLC) and PKC. The relevance of this response to adenosine cardioprotection must also be questioned, since although there is a report that ischemic preconditioning in the isolated rat heart is associated with the translocation of the δ-PKC isoform,[95] adenosine receptor antagonists do not block ischemic preconditioning in the rat.[58,60]

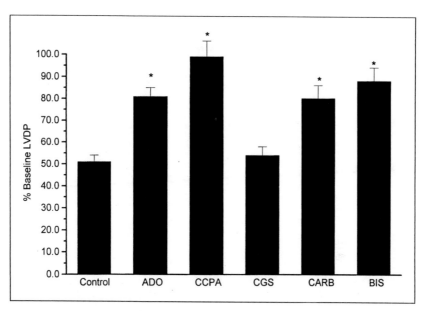

Fig. 4.7. Blockade of phorbol ester induced decreases in left ventricular developed pressure (LVDP) in constant flow perfused isolated rat hearts. In control hearts the phorbol ester phorbol 12-myristate 13-acetate (PMA) was infused for 30 minutes, and the hearts were monitored for an additional 30 minutes post-PMA. Results are expressed as % baseline LVDP. In additional PMA groups hearts were treated with adenosine (ADO), the ADO A_1 agonist CCPA, the ADO A_2 agonist CGS 21680, the acetylcholine analog carbachol (CARB) and the protein kinase C inhibitor bisindolylmaleimide (BIS). Agents were infused 20 minutes prior to PMA and throughout the remainder of the protocol. * $p < 0.05$ vs. control hearts.

There is other evidence that adenosine may modulate PKC, but in a negative manner. There are several reports that adenosine and adenosine agonists inhibit the effects of PKC activation in the cardiovascular system. Specifically, adenosine A_2 receptor stimulation inhibits phorbol ester induced PKC translocation and coronary artery constriction,[96] and reduces the deleterious effects of PKC activation in isolated lung preparations.[97] We have reported a similar antagonism of phorbol ester responses in isolated hearts by adenosine and adenosine receptor agonists.[98] Figure 4.7 illustrates the effects of adenosine and adenosine receptor agonists on phorbol ester-induced negative inotropy in isolated perfused rat hearts. The phorbol ester, phorbol 12-myristate 13-acetate (PMA,

10 nM), decreased left ventricular developed pressure (LVDP) by
$49 \pm 3\%$, an effect significantly reduced by the PKC inhibitor
bisindolylmaleimide (0.5 µM). Treatment with adenosine (100 µM),
the adenosine A_1 receptor agonist CCPA (0.1 µM), and the muscarinic agonist carbachol (0.5 µM) also greatly attenuated the negative inotropic effect of PMA. The adenosine A_2 analog CGS 21680
reduced the PMA-induced increase in coronary perfusion pressure but did not alter PMA's negative inotropic effect. We have
observed similar findings with other PKC activators and in the isolated rabbit heart. These results suggest that adenosine A_1 receptor
activation inhibits one or more isoforms of PKC or alters some
aspect of signal transduction distal to PKC. Since there is evidence
that PKC activity increases during ischemia,[99] it is possible that
adenosine, CCPA, and carbachol attenuate postischemic dysfunction by inhibiting the effects of ischemia-induced PKC activation.

Regardless of the specific signal transduction pathway(s) activated or inhibited by adenosine, there is evidence of the end-ef-
fectors that may be altered by adenosine in the ischemic-reperfused
heart. Since adenosine reduces catecholamine-induced increases
in intracellular calcium,[100] it is possible that adenosine may reduce
ischemia/reperfusion-induced calcium overload. This could occur
either via adenosine attenuation of catecholamine release and/or
the modulation of K_{ATP} channel activity. In fact Fralix and Murphy[101]
reported that isolated perfused rat hearts pretreated with adenosine exhibited less increase in intracellular calcium during global
ischemia and early reperfusion than untreated hearts. In addition
to reducing intracellular calcium overload during ischemia, adenosine may modulate intracellular calcium homeostasis during
reperfusion. There are reports that myocardial ischemia and reperfusion are associated with reductions in sarcoplasmic reticulum
(SR) Ca^{2+} uptake and decreased ryanodine-sensitive Ca^{2+} release
channel (RyR) function.[102,103] Preliminary evidence from our laboratory indicates that adenosine- and CCPA-mediated attenuation
of in vivo porcine myocardial stunning are associated with preservation of SR RyR function, but no alterations in stunning-induced
decreases in SR Ca^{2+} uptake.[104] Whether this effect, which was

present during both ischemia and reperfusion, has any causative role in adenosine improvement in postischemic systolic function or is merely a reflection of myocardial protection, remains to be determined.

Another mechanism of adenosine cardioprotection may be inhibition of oxygen free radical release and/or damage. Karmazyn and Cook[105] reported that adenosine and the A_1 receptor agonist PIA attenuated the negative inotropic effects of hydrogen peroxide in normoxic isolated perfused rat hearts. Xia et al[106] recently reported that the adenosine deaminase inhibitor EHNA, which increases endogenous adenosine levels, inhibited oxygen free radical generation (assessed by electron paramagnetic resonance) and improved postischemic ventricular function in isolated perfused rat hearts. It was proposed that this effect occurred by the inhibition of adenosine metabolism to hypoxanthine and xanthine, substrates for xanthine oxidase. However it remains to be determined whether this mechanism occurs in other species, including man, which have much less myocardial xanthine oxidase activity than the rat. There are other sources of oxygen free radicals, however, which can contribute to SR and contractile protein dysfunction by oxidation of critical sulfhydryl groups.

Adenosine reduction of irreversible injury may also occur via reduction of intracellular calcium overload and/or oxygen free radical damage. The reduction of ischemia-induced calcium overload is consistent with the results of studies in which adenosine and adenosine A_1 agonist pretreatments reduce infarct size presumably by exerting their protective effects during ischemia. The results of studies reporting reduced neutrophil accumulation in adenosine-treated irreversibly injured hearts are consistent with the hypothesis that adenosine may reduce infarct size, in part, via inhibiting neutrophil production of oxygen free radicals.

CARDIOPROTECTION AND ADENOSINE A_3 RECEPTORS

Another aspect of adenosine cardioprotection that has received recent attention is the potential involvement of adenosine A_3 receptors. Although there is evidence of cardiac adenosine A_3

mRNA and receptor expression in several species,[107,107] including man,[108,109] this hypothesis was initially based on the observations that DPCPX does not block ischemic preconditioning in the rabbit.[70,110] In addition, relatively high doses of 8-SPT are required to block ischemic preconditioning in the same species. Other support for this hypothesis is based on cardioprotection studies with the relatively new adenosine agonists APNEA and IB-MECA.[70,110,111] These agents have been reported to display A_3 receptor selectivity in membrane binding studies, but whether they are truly selective for the A_3 receptor at doses used in whole heart and isolated ventricular myocyte studies remains to be determined. The same concerns apply to the adenosine receptor antagonists DPCPX, 8-SPT and 8-(4-carboxyethylphenyl)-1,3-dipropylxanthine (A1433U), which have been used to provide pharmacological evidence of A_3 receptor involvement. In two recent studies using A1433U in rabbit myocardium, one group stated that this antagonist at 50 nM was selective for the A_1 receptor,[110] but another group reported that this same dose of A1433U did not alter the CCPA anti-adrenergic effect in rabbit ventricular myocytes,[112] an effect associated with A_1 receptor stimulation. Finally, there are several reports concluding that adenosine A_3 receptor activation is detrimental to the ischemic brain,[113,114] and that adenosine A_3 receptor activation may desensitize the cerebroprotective effects of adenosine A_1 receptor activation.[115] Thus the exact role, if any, of adenosine A_3 receptors in the ischemic heart, whether beneficial or detrimental, remains to be determined.

SUMMARY

Since human ventricular myocardium contains adenosine A_1 receptors that appear to be coupled to a pertussis toxin sensitive G protein,[116] it is likely that adenosine may be cardioprotective in the ischemic human heart. Three published studies have reported the safety and potential efficacy of exogenous adenosine in enhancing myocardial protection during open heart surgery in humans.[117-119] Despite these initial reports in humans and the extensive experimental evidence of adenosine's cardioprotective properties, much work remains to be done to elucidate its mechanism of action, optimal timing and dose. In addition more cardiac selective adenosine receptor agonists and adenosine potentiators need to be developed for both preclinical and clinical use.

REFERENCES

1. Belardinelli L, Linden J, Berne RM. The cardiac actions of adenosine. Prog Cardiovasc Dis 1989; 22:73-97.
2. Olsson RA, Pearson JD. Cardiovascular Purinoceptors. Physiol Rev 1990; 70(3):761-845.
3. Mubagwa K, Mullane K, Flameng W. Role of adenosine in the heart and circulation. Cardiovasc Res 1996; 32(5):797-813.
4. Reibel DK, Rovetto MJ. Myocardial ATP synthesis and mechanical function following oxygen deficiency. Am J Physiol 1978; 234: H620-H624.
5. Reibel DK, Rovetto MJ. Myocardial adenosine salvage rates and restoration of ATP content following ischemia. Am J Physiol 1979; 237:H247-H252.
6. Schrader J, Schutz W, Bardenheuer H. Role of S-adenosylhomocysteine hydrolase in adenosine metabolism in mammalian heart. Biochem J 1981; 196(1):65-70.
7. Achterberg PW, de Tombe PP, Harmsen E et al. Myocardial S-adenosylhomocysteine hydrolase is important for adenosine production during normoxia. Biochim Biophys Acta 1985; 840(3): 393-400.
8. Deussen A, Borst M, Kroll K et al. Formation of S-adenosylhomocysteine in the heart. II: A sensitive index for regional myocardial underperfusion. Circ Res 1988; 63(1):250-261.
9. Darvish A, Metting PJ. Purification and regulation of an AMP-specific cytosolic 5'-nucleotidase from dog heart. Am J Physiol 1993; 264(5 Pt 2):H1528-H1534.
10. Skladanowski AC, Smolenski RT, Tavenier M et al. Soluble forms of 5'-nucleotidase in rat and human heart. Am J Physiol 1996; 39(4):H1493-H1500.
11. Richardt G Blessing R, Schomig A. Cardiac noradrenaline release accelerates adenosine formation in the ischemic rat heart: role of neuronal noradrenaline carrier and adrenergic receptors. J Mol Cell Cardiol 1994; 26(10):1321-1328.
12. Dorheim TA, Hoffman A, Van Wylen DGL et al. Enhanced interstitial fluid adenosine attenuates myocardial stunning. Surgery 1991; 110:136-145.
13. Van Wylen DGL, Willis J, Sodhi J et al. Cardiac microdialysis to estimate interstitial adenosine and coronary blood flow. Am J Physiol 1990; 258:H1642-H1649.

14. Benson ES, Evans GT, Hallaway BE et al. Myocardial creatine phosphate and nucleotides in anoxic cardiac arrest and recovery. Am J Physiol 1961; 201:687-693.

15. Liu MS, Feinberg H. Incorporation of adenosine-8-^{14}C and inosine-8-^{14}C into rabbit heart adenine nucleotides. Am J Physiol 1971; 220:1242-1248.

16. Wiedmeier T, Rubio R Berne RM. Incorporation and turnover of adenosine-U-^{14}C in perfused guinea pig myocardium. Am J Physiol 1972; 223:51-54.

17. Silverman NA, Kohler J, Feinberg H et al. Beneficial metabolic effect of nucleoside augmentation on reperfusion injury following cardioplegic arrest. Chest 1983; 83:787-792.

18. Schubert T, Vetter H, Owen P et al. Adenosine cardioplegia. Adenosine versus potassium cardioplegia: effects on cardiac arrest and postischemic recovery in the isolated rat heart. J Thorac Cardiovasc Surg 1989; 98:1057-1065.

19. Ely SW, Mentzer RM Jr, Lasley RD et al. Functional and metabolic evidence of enhanced myocardial tolerance to ischemia and reperfusion with adenosine. J Thorac Cardiovasc Surg 1985; 90:549-556.

20. Wyatt DA, Ely SW, Lasley RD et al. Purine-enhanced asanguineous cardioplegia retards adenosine triphosphate degradation during ischemia and improves postischemic ventricular function. J Thorac Cardiovasc Surg 1989; 97:771-778.

21. Foker JE, Einzig S, Wang T. Adenosine metabolism and myocardial preservation. Consequences of adenosine catabolism on myocardial high-energy compounds and tissue blood flow. J Thorac Cardiovasc Surg 1980; 80:506-516.

22. Dhasmana JP, Digerness SB, Geckle JM et al. Effect of adenosine deaminase inhibitors on the heart's functional and biochemical recovery from ischemia: a study utilizing the isolated rat heart adapted to ^{31}P nuclear magnetic resonance. J Cardiovasc Pharmacol 1983; 5:1040-1047.

23. Koke JR, Fu LM, Sun D et al. Inhibitors of adenosine catabolism improve recovery of dog myocardium after ischemia. Mol Cell Biochem 1989; 86:107-113.

24. Van Belle H, Goossens F, Wynants J. Biochemical and functional effects of nucleoside transport inhibition in the isolated cat heart. J Mol Cell Cardiol 1989; 21:797-805.

25. Hudspeth DA, Williams MW, Zhao ZQ et al. Pentostatin-augmented interstitial adenosine prevents postcardioplegia injury in damaged hearts. Ann Thorac Surg 1994; 58:719-727.

26. Londos C, Wolff J. Two distinct adenosine-sensitive sites on adenylate cyclase. Proc Nat Acad Sci USA 1977; 74:5482-5486.

27. Van Calker D, Muller M, Hamprecht B. Adenosine regulates via two different types of receptors, the accumulation of cyclic AMP in cultured brain cells. J Neurochem 1979; 33:999-1005.

28. Olsson RA, Davis CJ, Khouri EM et al. Evidence for an adenosine receptor on the surface of dog coronary myocytes. Circ Res 1976; 39:93-98.
29. Collis MG. Evidence for an A_1-adenosine receptor in the guinea pig atrium. Br J Pharmacol 1983; 78:207-212.
30. Lasley RD, Rhee JW, Van Wylen DGL et al. Adenosine A_1 receptor mediated protection of the globally ischemic isolated rat heart. J Mol Cell Cardiol 1990; 22:39-47.
31. Lasley RD, Mentzer RM Jr. Adenosine improves the recovery of postischemic myocardial function via an adenosine A_1 receptor mechanism. Am J Physiol 1992; 263:H1460-H1465.
32. Yao Z, Gross GJ. Glibenclamide antagonizes adenosine A_1 receptor-mediated cardioprotection in stunned canine myocardium. Circulation 1993; 88:235-244.
33. Lasley RD, Noble MA, Konyn PJ et al. Different effects of an adenosine A_1 analogue and ischemic preconditioning in isolated rabbit hearts. Ann Thorac Surg 1995; 60:1698-1703.
34. Randhawa MPS Jr, Lasley RD, Mentzer RM Jr Salutary effects of exogenous adenosine on canine myocardial stunning in vivo. J Thorac Cardiovasc Surg 1995; 110:63-74.
35. Zhou Z, Bünger R, Lasley RD et al. Adenosine pretreatment increases cytosolic phosphorylation potential and attenuates postischemic cardiac dysfunction in swine. Surg Forum 1993; 44:249-252.
36. Sekili S, Jeroudi MO, Tang XL et al. Effect of adenosine on myocardial 'stunning' in the dog. Circ Res 1995; 76:82-94.
37. Yokota R, Fujiwara H, Miyamae M et al. Transient adenosine infusion before ischemia and reperfusion protects against metabolic damage in pig hearts. Am J Physiol 1995; 268:H1149-H1157.
38. Cave AC, Collis CS, Downey JM et al. Improved functional recovery by ischaemic preconditioning is not mediated by adenosine in the globally ischaemic isolated rat heart. Cardiovasc Res 1993; 27:663-668.
39. Olafsson B, Forman MB, Puett DW et al. Reduction of reperfusion injury in the canine preparation by intracoronary adenosine: importance of the endothelium and the no-reflow phenomenon. Circulation 1987; 76:1135-1145.
40. Pitarys CJ, Virmani R, Vildibill HD Jr et al. Reduction of myocardial reperfusion injury by intravenous adenosine administered during the early reperfusion period. Circulation 1991; 83:237-247.
41. Cronstein BN, Levin RI, Philips M et al. Neutrophil adherence to endothelium is enhanced via adenosine A_1 receptors and inhibited via adenosine A_2 receptors. J Immunol 1992; 148:2201-2206.
42. Soderback U, Sollevi A, Wallen NH et al. Anti-aggregatory effects of physiological concentrations of adenosine in human whole blood as assessed by filtragometry. Clin Sci 1991; 81:691-694.

43. Homeister JW, Hoff PT, Fletcher DD et al. Combined adenosine and lidocaine administration limits myocardial reperfusion injury. Circulation 1990; 82:595-608.

44. Vander Heide RS, Reimer KA. Effect of adenosine therapy at reperfusion on myocardial infarct size in dogs. Cardiovasc Res 1996; 31:711-718.

45. Norton ED, Jackson EK, Virmani R et al. Effect of intravenous adenosine on myocardial reperfusion injury in a model with low myocardial collateral blood flow. Am Heart J 1991; 122:1283-1291.

46. Goto M, Miura T, Iliodoromitis EK et al. Adenosine infusion during early reperfusion failed to limit myocardial infarct size in a collateral deficient species. Cardiovasc Res 1991; 25:943-949.

47. Norton ED, Jackson EK, Turner MB et al. The effects of intravenous infusions of selective adenosine A_1-receptor and A_2-receptor agonists on myocardial reperfusion injury. Am Heart J 1992; 123:332-338.

48. Thornton JD, Liu GS, Olsson RA et al. Intravenous pretreatment with A_1-selective adenosine analogues protects the heart against infarction. Circulation 1992; 85:659-665.

49. Schlack W, Schafer M, Uebing A et al. Adenosine A_2-receptor activation at reperfusion reduces infract size and improves myocardial wall function in dog heart. J Cardiovasc Pharmacol 1993; 22:89-96.

50. Jordan JE, Zhao ZQ, Sato H et al. Adenosine A_2 receptor activation attenuates reperfusion injury by inhibiting neutrophil accumulation, superoxide generation and coronary endothelial adherence. J Pharmacol Exp Ther 1997; 280(1):301-309.

51. Salmon JE, Cronstein BN. Fcg receptor-mediated functions in neutrophils are modulated by adenosine receptor occupancy. Adenosine A_1 receptors are stimulatory and A_2 receptors are inhibitory. J Immunol 1990; 145:2235-2240.

52. Schwartz LM, Raschke P, Becker BF et al. Adenosine contributes to neutrophil-mediated loss of myocardial function in post-ischemic guinea-pig hearts. J Mol Cell Cardiol 1993; 25:927-938.

53. Raschke P, Becker BF. Adenosine and PAF dependent mechanisms lead to myocardial reperfusion injury by neutrophils after brief ischaemia. Cardiovasc Res 1995; 29:569-576.

54. Auchampach JA, Gross GJ. Adenosine A_1 receptors, K_{ATP} channels, and ischemic preconditioning in dogs. Am J Physiol 1993; 264: H1327-H1336.

55. Van Winkle DM, Chien GL, Wolff RA et al. Cardioprotection provided by adenosine receptor activation is abolished by blockade of the K_{ATP} channel. Am J Physiol 1994 Feb;266(2 Pt 2):H829-H839.

56. Toombs CF, McGee S, Johnston WE et al. Myocardial protective effects of adenosine. Infarct size reduction with pretreatment and continued receptor stimulation during ischemia. Circulation 1992 Sep; 86(3):986-994

57. Lasley RD, Konyn PJ, Hegge JO et al. Effects of ischemic and adenosine preconditioning on interstitial fluid adenosine and myocardial infarct size. Am J Physiology 1995; 269:H1460-H1466.

58. Li Y, Kloner RA. The cardioprotective effects of ischemic preconditioning are not mediated by adenosine receptors in rat hearts. Circulation 1993; 87:1642-1648.

59. Li Y, Kloner RA. Adenosine deaminase inhibition is not cardioprotective in the rat. Am Heart J 1993; 126(6):1293-1298.

60. Liu Y, Downey JM. Ischemic preconditioning protects against infarction in rat hearts. Am J Physiol 1992; 263:H1107-H1112.

61. Liu GS, Jacobson KA, Downey JM. An irreversible A_1 selective adenosine agonist preconditions rabbit heart. Can J Cardiol 1996; 12:517-521.

62. Liu GS, Thornton J, Van Winkle DM et al. Protection against infarction afforded by preconditioning is mediated by A_1 adenosine receptors in rabbit heart. Circulation 1991; 84:350-356.

63. Murry CE, Jennings RB, Reimer KA. Preconditioning with ischemia: a delay of lethal cell injury in ischemic myocardium. Circulation 1986; 74:1124-1136.

64. Van Wylen DG. Effect of ischemic preconditioning on interstitial purine metabolite and lactate accumulation during myocardial ischemia. Circulation 1994; 89(5):2283-2289.

65. Miura T, Ogawa T, Iwamoto T et al. Dipyridamole potentiates the myocardial infarct size-limiting effect of ischemic preconditioning. Circulation 1992; 86(3):979-985.

66. Itoya M, Miura T, Sakamoto J et al. Nucleoside transport inhibitors enhance the infarct size-limiting effect of ischemic preconditioning. J Cardiovasc Pharmacol 1994; 24(5):846-852.

67. Silva PH, Dillon D, Van Wylen DG. Adenosine deaminase inhibition augments interstitial adenosine but does not attenuate myocardial infarction. Cardiovasc Res 1995; 29:616-623.

68. Martin BJ, Lasley RD, Mentzer RM Jr. Infarct size reduction with the nucleoside transport inhibitor R75231 in swine. Am J Physiol 1997; 272:H1857-H1865.

69. Zhao ZQ, McGee S, Nakanishi K et al. Receptor-mediated cardioprotective effects of endogenous adenosine are exerted primarily during reperfusion after coronary occlusion in the rabbit. Circulation 1993; 88(2):709-719.

70. Liu GS, Richards SC, Olsson RA et al. Evidence that the adenosine A_3 receptor may mediate the protection afforded by preconditioning in the isolated rabbit heart. Cardiovasc Res 1994; 28(7):1057-1061.

71. Lasley RD, Anderson GM, Mentzer RM Jr. Ischemic and hypoxic preconditioning enhance postischemic recovery of function in the rat heart. Cardiovasc Res 1993; 27:565-570.

72. Ganote CE, Armstrong S, Downey JM. Adenosine and A_1 selective agonists offer minimal protection against ischaemic injury to isolated rat cardiomyocytes. Cardiovasc Res 1993; 27:1670-1676.

73. Belardinelli L, Shryock JC, Song Y et al. Ionic basis of the electrophysiological actions of adenosine on cardiomyocytes. FASEB J 1995; 9:359-365.

74. Lasley RD, Mentzer RM Jr. Pertussis toxin blocks adenosine A_1 receptor mediated protection of the ischemic rat heart. J Mol Cell Cardiol 1993; 25:815-821.

75. Thornton JD, Liu GS, Downey JM. Pretreatment with pertussis toxin blocks the protective effects of preconditioning: evidence for a G-protein mechanism. J Mol Cell Cardiol 1993; 25(3):311-320.

76. Wennmalm M, Fredholm BB, Hedqvist P. Adenosine as a modulator of sympathetic nerve-stimulation-induced release of noradrenaline from the isolated rabbit heart. Acta Physiol Scand 1988; 132:487-494.

77. Dobson JG, Fenton RA, Romano FD. The antiadrenergic actions of adenosine in the heart. In: Gerlach, E and Becker BF, eds. Topics and Perspectives in Adenosine Research. Berlin, Heidelberg: Springer-Verlag, 1987:356-368.

78. Richardt G, Waas W, Kranzhofer R et al. Adenosine inhibits exocytotic release of endogenous noradrenaline in rat heart: a protective mechanism in early myocardial ischemia. Circ Res 1987; 61:117-123.

79. Rynning SE, Brunvand H, Birkeland S et al. Endogenous adenosine attenuates myocardial stunning by antiadrenergic effects exerted during ischemia and not during reperfusion. J Cardiovasc Pharmacol 1995; 25:432-439.

80. Fenton RA, Galeckas KJ, Dobson JG Jr. Endogenous adenosine reduces depression of cardiac function induced by beta-adrenergic stimulation during low flow perfusion. J Mol Cell Cardiol 1995; 27:2373-2383.

81. Toombs CF, McGee DS, Johnston WE et al. Protection from ischaemic-reperfusion injury with adenosine pretreatment is reversed by inhibition of ATP sensitive potassium channels. Cardiovasc Res 1993; 27(4):623-629.

82. Yao ZH, Mizumura T, Mei DA et al. K-ATP channels and memory of ischemic preconditioning in dogs—synergism between adenosine and K_{ATP} channels. Am J Physiol 1997; 41:H334-H342.

83. Kirsch GE, Codina J Birnbaumer L et al. Coupling of ATP-sensitive K^+ channels to A_1 receptors by G proteins in rat ventricular myocytes. Am J Physiol 1990; 259(3 Pt 2):H820-H826.

84. Ito H, Vereecke J, Carmeliet E. Mode of regulation of G protein of the ATP-sensitive K^+ channel in guinea-pig ventricular cell membrane. J Physiol (London) 1994; 478:101-107.

85. Liu YG, Gao WD, O'ourke B et al. Synergistic modulation of ATP-sensitive K^+ currents by protein kinase C and adenosine—implications for ischemic preconditioning. Circ Res 1996; 78:443-454.

86. Babenko AP, Vassort G. Purinergic facilitation of ATP-sensitive potassium current in rat ventricular myocytes. Brit J Pharmacol 1997; 120(4):631-638.

87. Grover GJ, Baird AJ, Sleph PG. Lack of a pharmacologic interaction between ATP-sensitive potassium channels and adenosine A_1 receptors in ischemic rat hearts. Cardiovasc Res 1996; 31:511-517.

88. Urquhart RA, Broadley KJ. The indirect negative inotropic effects of the P1-receptor agonist L-phenylisopropyladenosine, in guinea pig isolated cardiac preparations: comparison with cromakalim. Can J Physiol 1992; 70:910-915.

89. Song Y, Srinivas M, Belardinelli L. Nonspecific inhibition of adenosine-activated K^+ current by glibenclamide in guinea pig atrial myocytes. Am J Physiol 1996; 271(6 Pt 2):H2430-H2437.

90. Nakhostine N, Lamontagne D. Adenosine contributes to hypoxia-induced vasodilation through ATP-sensitive K^+ channel activation. AmJ Physiol 1993; 265:H1289-H1293.

91. Ytrehus K, Liu Y, Downey JM. Preconditioning protects ischemic rabbit heart by protein kinase C activation. Am J Physiol 1994; 266:H1145-H1152.

92. Sakamoto J, Miura T, Goto M et al. Limitation of myocardial infarct size by adenosine A_1 receptor activation is abolished by protein kinase C inhibitors in the rabbit. Cardiovasc Res 1995; 29(5):682-688.

93. Armstrong SC, Hoover DB, Delacey MH et al. Translocation of PKC, protein phosphatase inhibition and preconditioning of rabbit cardiomyocytes. J Mol Cell Cardiol 1996; 28:1479-1492.

94. Henry P, Demolombe S, Puceat M et al. Adenosine A_1 stimulation activates d protein kinase C in rat ventricular myocytes. Circ Res 1996; 78:161-165.

95. Mitchell MB, Meng X, Ao L et al. Preconditioning of isolated rat heart is mediated by protein kinase C. Circ Res 1995; 76:73-81.

96. Cushing DJ, Makujina SR, Sabouni MH et al. Protein kinase C and phospholipase C in adenosine receptor-mediated relaxation in coronary artery. Am J Physiol 1991; 261 (Heart Circ. Physiol. 30): H1848-H1854.

97. Adkins WK, Barnard JW, Moore TM et al. Adenosine prevents PMA-induced lung injury via an A_2 receptor mechanism. J Appl Physiol 1993; 74:982-988.

98. Lasley RD, Noble MA, Paulsen KL et al. Adenosine attenuates phorbol ester-induced negative inotropic and vasoconstrictive effects in the isolated rat heart. Am J Physiol 1994; 266:H2159-H2166.

99. Yoshida K, Hirata T, Akita Y et al. Translocation of protein kinase C-α, δ and ϵ isoforms in ischemic rat heart. Biochim Biophys Acta 1996; 1317:36-44.

100. Fenton RA, Moore EDW, Fay FS et al. Adenosine reduces the Ca^{2+} transients of isoproterenol- stimulated rat ventricular myocytes. Am J Physiol 1991; 261:C1107-C1114.

101. Fralix TA, Murphy E, London RE et al. Protective effects of adenosine in the perfused rat heart: changes in metabolism and intracellular ion homeostasis. Am J Physiol 1993; 264:C986-C994.

102. Wu QY, Feher J. Effect of ischemia and ischemia-reperfusion on ryanodine binding and Ca^{2+} uptake of cardiac sarcoplasmic reticulum. J Mol Cell Cardiol 1995; 27:1965-1975.

103. Zucchi R, Ronca-Testoni S, Yu G et al. Effects of ischemia and reperfusion on cardiac ryanodine receptors sarcoplasmic reticulum Ca^{2+} channels. Circ Res 1994; 74:271-280.

104. Valdivia CR, Lasley RD, Hegge JO et al. Adenosine pretreatment prevents myocardial stunning-induced reduction of ryanodine receptor function. Circulation 1996; 94(Suppl I):I-185.

105. Karmazyn M, Cook MA. Adenosine A_1 receptor activation attenuates cardiac injury produced by hydrogen peroxide. Circ Res 1992; 71:1101-1110.

106. Xia Y, Khatchikian G, Zweier JL. Adenosine deaminase inhibition prevents free radical-mediated injury in the postischemic heart. J Biol Chem 1996; 271:10096-10102.

107. Hill RJ, Oleynek JJ, Hoth CF et al. Cloning, expression and pharmacological characterization of rabbit adenosine A_1 and A_3 receptors. J Pharmacol Exp Ther 1997; 280(1):122-128.

108. Linden J. Cloned adenosine A_3 receptors: pharmacological properties, species differences and receptor functions. Trends Pharmacol Sci 1994; 15:298-306.

109. Salvatore CA, Jacobson MA, Taylor HE et al. Molecular cloning and characterization of human A_3 adenosine receptor. Proc Natl Acad Sci USA 1993; 90:10365-10369.

110. Armstrong S, Ganote CE. Adenosine receptor specificity in preconditioning of isolated rabbit cardiomyocytes: evidence of A_3 receptor involvement. Cardiovasc Res 1994;28(7):1049-1056.

111. Tracey WR, Magee W, Masamune H et al. Selective adenosine A_3 receptor stimulation reduces ischemic myocardial injury in the rabbit heart. Cardiovasc Res 1997; 33:410-415.

112. Wang JX, Drake L, Sajjadi F et al. Dual activation of adenosine A_1 and A_3 receptors mediates preconditioning of isolated cardiac myocytes. Eur J Pharmacol 1997; 320:241-248.

113. Von Lubitz DK, Lin RC, Popik P et al. Adenosine A_3 receptor stimulation and cerebral ischemia. Eur J Pharmacol 1994; 263(1-2):59-67.

114. Abbracchio MP, Brambilla R, Ceruti S et al. G protein-dependent activation of phospholipase C by adenosine A_3 receptors in rat brain. Mol Pharmacol 1995; 8(6):1038-1045.

115. Dunwiddie TV, Diao LH, Kim HO et al. Activation of hippocampal adenosine A_3 receptors produces a desensitization of A(1) receptor-mediated responses in rat hippocampus. J Neurosci 1997; 17(2):607-614.

116. Bohm M, Pieske B, Ungerer M et al. Characterization of A_1 adenosine receptors in atrial and ventricular myocardium from diseased human hearts. Circ Res 1989; 65:1201-1211.

117. Lee HT, Lafaro RJ, Reed GE. Pretreatment of human myocardium with adenosine during open heart surgery. J Cardiac Surg 1995; 10:665-676.

118. Fremes SE, Levy SL, Christakis GT et al. Phase 1 human trial of adenosine-potassium cardioplegia. Circulation 1996; 94(Suppl II): 370-375.

119. Mentzer RM, Rahko PS, Molina-Viamonte V et al. Safety, tolerance, and efficacy of adenosine as an additive to blood cardioplegia in humans during coronary artery bypass surgery. Am J Cardiol 1997; (in press).

CHAPTER 5

The Role of Adenosine in the Regulation of Cardiac Function: The Effect of Aging

Guoping Cai, Hoau-Yan Wang and Eitan Friedman

INTRODUCTION

ROLE OF ADENOSINE IN THE REGULATION OF CARDIAC FUNCTION

A denosine, an endogenous metabolite, plays a crucial role in the regulation of cardiac energy metabolism and physiological functions. The first actions of extracellular adenosine on the heart were described as a negative chronotropic action on sinus node automaticity, a negative dromotropic action on atrio-ventricular (AV) nodal conduction and lusitropic action on coronary blood vessels.[1] The latter action gave rise and was the basis of the proposed "adenosine hypothesis",[2,3] according to which, the reduction in myocardial oxygen supply and/or increase in myocardial workload (i.e., oxygen demand) promotes enhanced adenosine production and release, which leads to an increase in coronary blood flow. Later observations showed that adenosine also depresses myocardial contractility as well as inhibits ventricular automaticity.[4-7] These

Effects of Extracellular Adenosine and ATP on Cardiomyocytes,
edited by Amir Pelleg and Luiz Belardinelli. © 1998 R.G. Landes Company.

physiological effects result in an increase in oxygen supply and a reduction in oxygen demand, and thereby maintain energy balance and protect the myocardium from injury during acute myocardial ischemia.

On the other hand, the antiadrenergic effects of adenosine are well recognized.[4,8] Schrader et al[9] reported that adenosine produced a dose-dependent inhibition of isoproterenol-stimulated inotropic action, which was mediated by the inhibitory effect of adenosine on the adenylyl cyclase system.[10,11] Adenosine also antagonizes the positive chronotropic effect which is induced by catecholamine.[12] In addition, adenosine attenuates presynaptic norepinephrine release from adrenergic nerve terminals innervating the heart.[13] Thus, adenosine, through its presynaptic and postsynaptic effects, acts functionally as an inhibitor of catecholamine stimulation, thereby protecting the heart from excessive sympathetic stimulation.

ADENOSINE RECEPTOR SUBTYPES
AND TRANSMEMBRANE SIGNALING

Adenosine's signal transduction is discussed in detail in another chapter in this volume. Therefore, only the aspects relevant to the issue of aging are briefly discussed below. Adenosine is produced intracellularly and transported into the interstitial space, where it exerts its physiological action by binding to cell surface receptors. Burnstock[14] originally proposed a new class of receptors, purinoceptors, different from cholinergic and adrenergic receptors, which mediate the actions of extracellular adenosine and ATP. These receptors were called P_1 and P_2, respectively.[15] Adenosine receptors, i.e., P_1 receptors, were classified into the A_1 and A_2 subtypes mediating the inhibition and stimulation of adenylyl cyclase, respectively.[16,17] Additional adenosine receptor subtypes were identified using various selective analogs;[5,18] four adenosine receptor subtypes have been cloned to date, i.e., A_1, A_{2a}, A_{2b} and A_3.[19,20] Structurally, these adenosine receptor subtypes have seven membrane-spanning segments which is a characteristic they share with all G protein-coupled cell surface receptors.[5,20,21] In addition to the inhibition of adenylyl cyclase, A_1-adenosine receptors are also coupled to a variety of other effector systems. It is well documented that adenosine activates K^+ channels in cardiac myocytes acting via A_1-adenosine receptors which are coupled to a pertussis toxin-sensitive G protein.[4,22] Many of the electrophysiologic effects of adenosine appear to be related to alteration in K^+ conductance rather than alteration in cyclic AMP level. Other effectors that have been reported to be modulated by A_1-adenosine receptors include phospholipase C, Ca^{2+} channel, phospholipase A_2, glucose transporter and the Na^+/Ca^{2+} exchanger.[5]

CHANGES IN CARDIOVASCULAR FUNCTION DURING AGING

Aging is accompanied by a variety of changes in cardiovascular function and metabolism.[23] The most significant hemodynamic change seen with aging is increased blood pressure associated with either unchanged or decreased heart rate. The increase in blood pressure is related to an increase in peripheral vascular resistance resulting from either altered vascular responsiveness to vasoactive factors or structural remodeling of blood vessels. As a consequence of this elevated cardiac afterload, the heart increases its contractility so as to maintain cardiac output. Thus, in the elderly, the cardiovascular system continues to function reasonably well at rest but cardiac reserve is greatly reduced. Aging is also associated with increased incidence of cardiovascular diseases including coronary artery disease, heart failure, hypertension and postural hypotension, although it is difficult to separate purely age-related changes from underlying pathological changes. Since adenosine is an important regulator of cardiac function and has been identified as an endogenous cardioprotective substance, this chapter reviews age-related changes in the metabolism and function of adenosine in the heart, focusing on myocardial A_1-adenosine receptor function during aging.

AGE-RELATED CHANGES IN CARDIAC EFFECTS
OF ADENOSINE: A BRIEF OVERVIEW

PRODUCTION AND METABOLISM OF ADENOSINE

Adenosine is formed by dephosphorylation of adenosine monophosphate (AMP), catalyzed by the plasma membrane-bound or cytosolic 5'-nucleotidase (ATP pathway). Intracellular degradation of 5-adenosylhomocysteine (SAH), catalyzed by S-adenosylhomocysteine hydrolase, also leads to the formation of adenosine (SAH pathway). Adenosine can be either deaminated by adenosine deaminase to inosine or phosphorylated by adenosine kinase to AMP. Since both adenosine deaminase and adenosine kinase are cytosolic enzymes, adenosine in the interstitium must first be taken up by the cells, by either simple or facilitated diffusion, before it is metabolized (see chapter 2).

Interstitial adenosine level in the heart, measured in coronary effluent or in cardiac microdialysate, is increased with age.[24-26] Increases in adenosine levels were found in adipocytes obtained from old rats[27] and in senescent fibroblasts.[28] Age-related increase in interstitial adenosine was also observed during cardiac maturation.[29] While the activity of adenosine deaminase is increased with age in human skin fibroblasts[30] and adipocytes,[31] adenosine metabolism is decreased in lymphocytes taken from aged mice.[32,33] Nevertheless, it was reported that the activities of 5'-nucleotidase and adenosine deaminase were comparable in the hearts of young and old rats.[25] These results, therefore, may suggest that the activities of these enzymes are not likely to be responsible for the increase in adenosine production in the aged heart. Possible changes in the activities of SAH hydrolase and adenosine kinase need to be quantitated before age-dependent changes in the metabolic pathways of adenosine are fully determined. In addition, impaired transport of adenosine into cells could account for the increase in adenosine concentration in myocardial interstitial space of the aged heart. However, another study suggested that increase in adenosine efflux in the heart during aging was not due to altered uptake or catabolism of adenosine but was associated with a two-fold higher cytosolic AMP concentration despite an unchanged energy state.[26]

Since adenosine attenuates the effects of β-adrenergic receptor stimulation, the increase in adenosine level during aging has been suggested to mediate the age-related decline in the apparent responsiveness of β receptors to their agonists.[24,25] However, a recent study showed that no age-related change in adenosine content could be found in the hearts of aged rats when myocardial adenosine level was directly measured by high performance liquid chromatography.[34] The plasma level of adenosine in control human subjects was also shown to be unchanged with age.[35] Thus, the effect of age on extracellular adenosine levels in the heart is still to be conclusively determined. In addition, although adenosine production was increased in both young and aged hearts during ischemia, myocardial adenosine content was 50% lower in the aged heart after 5 and 10 minutes of ischemia and remained depressed for up to 25 minutes, suggesting that a diminished increase in adenosine during ischemia may contribute to the greater susceptibility of the aged heart to ischemic damage.[34]

CARDIAC ELECTROPHYSIOLOGIC EFFECTS OF ADENOSINE

The electrophysiologic effects of adenosine include the negative chronotropic effect on the sinoatrial (SA) node and the negative dromotropic effect on the atrio-ventricular (AV) node. These effects of adenosine are mediated by the activation of K_{Ado} channels and an increase in outward K^+ current as well as attenuation of Ca^{2+} current.[36,37] The former hyperpolarizes the cell membrane, decreases the rate of pacemaker cell depolarization and depresses the action potential in AV nodal cells. In addition, adenosine attenuates catecholamine-enhanced pacemaker I_f and inward Ca^{2+} currents.[36-38]

Age-related changes in the electrophysiologic effects of adenosine in the heart are less clear. In the isolated perfused neonate rabbit heart, exogenous adenosine produced a smaller dose-dependent increase in A-H interval than in adult hearts.[39] The ability of adenosine to slow spontaneous heart rate became greater during the course of organ culture of young embryonic chick hearts.[40] Thus, developmental maturation in adenosine-mediated regulation of electrophysiologic functions seems to occur in superventricular tissues. Similarly, the negative chronotropic effect of adenosine was greater in right atria of old than young rats.[41,42] The adenosine-induced bradycardia, presumably an A_1-adenosine receptor mediated effect, was 10-fold more pronounced in aged than in young adult hearts.[26]

Opposite results were also obtained. Specifically, the isolated sinus node obtained from adult rats was more sensitive to adenosine than that from old rats.[43] This could have been due to an alteration in the intrinsic activity of SA nodal cells, since Ca^{2+}-induced effects were also altered.[43] A recent in vivo study demonstrated that during isoproterenol challenge the negative chronotropic action of adenosine, determined as prolongation of sinus cycle length, was decreased in old Fischer 344 rats (24 month old) compared to adult animals (6 month old), although there was no difference between the action of adenosine in the adult vs. old rats under basal conditions.[44] Furthermore, under conditions of acute transient global myocardial hypoxia in the presence of isoproterenol, atrial fibrillation was induced by rapid atrial pacing in one of six adult rats but in all six old rats.[44] This age-related difference in induction of atrial fibrillation was diminished when the young animals were pretreated with A_1-adenosine receptor antagonists.[44] The results of these studies indicate that the efficacy of the anti-β adrenergic action of exogenous or endogenous adenosine decreases with age, as manifested by a reduction in negative chronotropic action of adenosine and an increase in vulnerability to atrial fibrillation during isoproterenol infusion.

Clinical studies have shown that the dose-response relationship of the effects of adenosine on heart rate did not differ significantly between young and old individuals, indicating that age does not modulate the direct actions of adenosine in humans.[45] However, a higher dose of isoproterenol was required to increase heart rate in old vs. young human subjects, and this age-related difference was abolished by pretreatment with the adenosine receptor antagonist, theophylline, implying a higher endogenous adenosine activity in hearts of old vs. young individuals.[46] The effect of aging on the direct and indirect, i.e., anti-β adrenergic, actions of adenosine in humans still needs to be conclusively determined.

ADENOSINE-MEDIATED INHIBITION
OF MYOCARDIAL CONTRACTILITY

In atrial myocytes, the principal effect of adenosine is to activate K_{Ado} channels. This results in the induction of an outward K^+ current which shortens the action potential duration and eventually decreases atrial contractility.[4] Adenosine, indirectly through G_i protein mediated inhibition of adenylyl cyclase, decreases intracellular cyclic AMP level, and thereby attenuates the stimulation of L-type Ca^{2+} channels induced by β-adrenergic receptor agonists.[4,11] In contrast, it is still widely believed that adenosine does not exert a direct effect in mammalian ventricular tissue, though distinct K_{Ado} channels have been identified in ventricular myocytes of various species, including human[7,47,48] and ferret[49] ventricular myocytes. However, adenosine effectively antagonizes the positive inotropic effects of β-adrenergic agonists on ventricular myocardium in vitro.[4,50] This antiadrenergic effect of adenosine in ventricular myocardium could be due to its antagonism of catecholamine-stimulated adenylyl cyclase activity and L-type Ca^{2+} current.[9,37,51,52] It has been suggested that adenosine may also interfere with the interaction of adrenergic receptors with their agonists by reducing the high affinity binding capacity of the β-adrenergic receptor.[53] However, the similarity in adenosine-mediated attenuation of the effects of isoproterenol and forskolin argues against such a mechanism.

The negative inotropic effects of adenosine are the result of its interaction with A_1-adenosine receptors. The function of the A_1-adenosine receptor is quite well developed in the immature heart since this receptor is present in great numbers and is effectively coupled to its effector.[54] In the isolated perfused heart preparation, the dose of adenosine required to dilate the heart was lower in immature vs. mature hearts.[55] Furthermore, adenosine did not significantly inhibit isoproterenol-stimulated Ca^{2+} current and cyclic

AMP levels in adult rabbit ventricular myocytes, while it completely blocked the isoproterenol-induced Ca^{2+} current and diminished isoproterenol-stimulated cyclic AMP level in newborn animals.[56] These adenosine-mediated effects in immature hearts seemed to be mediated primarily through the G_{i3} protein pathway since a large amount of $G_{\alpha i3}$ protein is found in the immature heart and the effect of adenosine on Ca^{2+} current was partially blocked by the synthetic decapetide corresponding to the C-terminal sequence of $G_{\alpha i3}$ protein.[56] However, results derived from embryonic chick hearts failed to show a difference in adenosine-mediated inhibition of isoproterenol-induced accumulation of cyclic AMP during the development of the heart.[57] In a recent study, isoproterenol produced a smaller increase in venous adenosine concentration and the isoproterenol-elicited contractile response was not altered by pretreatment with an A_1-adenosine receptor antagonist in immature hearts, implying that A_1-adenosine receptor function was not well developed in the immature heart.[58]

It is well known that the β-adrenoceptor stimulated myocardial contractile response is decreased during aging. This was suggested to be due to increased adenosine release during catecholamine stimulation.[24,25] In addition, a recent study showed that adenosine-mediated inhibition of isoproterenol-stimulated adenylyl cyclase activity was enhanced in aged rats and this was accompanied by a 50% increase in A_1-adenosine receptor density.[59] Thus, it has been hypothesized that increased production of adenosine, higher A_1-adenosine receptor density and enhanced A_1-adenosine receptor function contribute to the reduction in isoproterenol-elicited contractile response observed in aged hearts. However, others have noted an unchanged[41] or reduced[42] negative inotropic effect of adenosine in atria obtained from aged compared to young adult rats. Recent extensive studies performed in our laboratory in Fischer 344 rats using a more selective A_1-adenosine receptor agonist, gave evidence for an age-related deterioration of A_1-adenosine receptor function in both ventricular and atrial tissues.[60,61] Supporting data were obtained in an in vivo model.[44] These findings are reviewed below.

REGULATION OF CORONARY BLOOD FLOW BY ADENOSINE

Adenosine, acting via A_2-adenosine receptors produces a potent vasodilation in the majority of vascular beds including the coronary blood vessels. The proposed mechanisms for this effect include elevation of cyclic AMP by stimulation of adenylyl cyclase, increased release of nitric oxide through the guanylyl cyclase pathway, inhibition of inositol phosphate accumulation, depression of Ca^{2+}-dependent action potential and inhibition of Ca^{2+} uptake.[36,62,63] It was also recently suggested that adenosine-induced vasodilation is mediated by the activation of K_{ATP} channels in vascular smooth muscle cells.[64,65] However, which adenosine receptor subtype and whether G protein are involved in the activation of the K_{ATP} channel is still controversial.[66-69]

Age-related changes in the responses to adenosine have been observed in a variety of vascular preparations. Adenosine-induced relaxation of mesenteric arteries,[70] muscle microvasculature[71] and aortic rings[26] decreased with age. Aortic rings obtained from immature guinea pigs displayed a more sensitive and greater response to adenosine than those obtained from mature animals.[55] However, it was found that there was an age-related increase in responsiveness to exogenous adenosine and adenosine was not as effective a vasodilator in coronary and internal arteries of the newborn as it was in that of the adult, suggesting that the development of A_2-adenosine receptor function accompanied maturation of these vessels.[72,73] Another study showed that adenosine-mediated relaxation of coronary arteries in beagle dogs did not differ during aging.[74] Thus, adenosine-mediated vascular relaxation as a function of age may differ from vessel to vessel and from one species to another.

MODULATION OF PRESYNAPTIC NEUROTRANSMITTER RELEASE BY ADENOSINE

In addition to its postsynaptic effects, adenosine was also shown to inhibit norepinephrine release from nerve terminals.[75-77] The presynaptic release of norepinephrine is significantly increased in the presence of adenosine receptor antagonists, theophylline or

8-phenyltheophylline.[78,79] This apparent regulatory effect of adenosine on norepinephrine release is mediated by presynaptic A_1-adenosine receptor.[80-82] Thus, sufficient accumulations of extracellular adenosine during early stage of ischemia may effectively reduce presynaptic norepinephrine release and thereby provide another way of protecting the myocardium from injury caused by excessive β-adrenoceptor stimulation, increased cardiac contractility and reduced O_2 supply. In cardiac synaptosomes prepared from Fischer 344 rats, selective A_1-adenosine receptor stimulation with the agonist, N^6-p-sulfophenyladenosine (SPA), inhibited potassium-induced norepinephrine release.[83] This effect was markedly attenuated by the A_1-adenosine receptor antagonists, 8-p-sulfophenyl-theophylline (8-SPT) but was not modified by an A_2-adenosine receptor agonist.[83] The A_1-adenosine receptor-mediated inhibitory effect on presynaptic norepinephrine release was significantly greater in adult rats (6 month old) compared to old animals (24 month old).[83] These results suggest that A_1-adenosine receptor-mediated modulation of presynaptic norepinephrine release declines with age, which may, at least in part, explain the diminished antiadrenergic protective action of adenosine in the aged heart.

AGE-RELATED CHANGES IN THE COUPLING OF A_1-ADENOSINE RECEPTORS TO G PROTEINS

A_1 ADENOSINE RECEPTOR-MEDIATED INHIBITION OF ADENYLYL CYCLASE ACTIVITY

The A_1-adenosine receptor is the main adenosine receptor subtype in the ventricular myocardium, which mediates the negative inotropic action of adenosine.[4,5,7,84] The activation of this receptor, coupled to a pertussis-toxin sensitive G protein, is known to inhibit adenylyl cyclase.[84-86] Thus, the activation of A_1-adenosine receptors, under the condition of stimulated adenylyl cyclase, decreases intracellular cyclic AMP levels and reduces myocardial contractility. Since A_1-adenosine receptors in ventricular myocardium do not affect the K_{Ado} channel, adenosine causes no discernible change in basal myocardial contractile activity.[4] Nevertheless, adenosine effectively inhibits adenylyl cyclase which is stimulated by either β-adrenoceptor activation or by the catalytic stimulator of adenylyl cyclase, forskolin.[51,87-89]

In order to determine the age-related changes in myocardial A_1-adenosine receptor function, receptor-mediated inhibition of adenylyl cyclase activity was examined in ventricular membranes obtained from adult (6 month old) and old (24 month old) Fischer 344 rats.[60,61] The selective A_1-adenosine receptor agonists, SPA or N^6-cyclopentladenosine (CPA), dose-dependently inhibited either isoproterenol- or forskolin-stimulated adenylyl cyclase activity in ventricular membranes. This effect of the agonists was blocked by the A_1-adenosine receptor antagonist, 8-SPT. A_1-adenosine receptor-mediated inhibition of adenylyl cyclase activity was greatly attenuated in old rats compared to adult animals, suggesting an age-related impairment in the function of cardiac ventricular A_1-adenosine receptors. Our data appear to be inconsistent with previous results.[59] This discrepancy may be related to differences in a number of experimental details employed in the two studies. Firstly, two different adenosine analogs were used in these studies. In our experiments, a selective A_1-adenosine receptor agonist, SPA, was used so as not to confound the results obtained when using the less selective agonist, N^6-(2-phenylisopropyl) adenosine (PIA). In embryonic chick myocytes, PIA acts at more than one receptor to inhibit adenylyl cyclase.[90] The A_1/A_2 receptor selectivity ratio for PIA is not as favorable as it is for CPA or SPA.[91,92] In addition, PIA can affect the high affinity binding to purinergic receptors,[53] which might influence the ability of low concentrations of isoproterenol to stimulate adenylyl cyclase. Furthermore, the study by Romano and Dobson[59] has defined 18 month old rats as aged, and has included a mixture of both Sprague-Dawley and Fischer strains in their experimental groups. These two rat strains are known to have different life spans and to differ in many biological aspects. These facts make it difficult to compare our experimental results with those previously obtained by Romano and Dobson.[59]

The activation of muscarinic cholinergic receptors, via a G_i protein, also inhibits the activity of adenylyl cyclase. We therefore tested the specificity of the age-related reduction in myocardial A_1-adenosine receptor function. Consistent with previous studies,[93-95] we noted no age-related decline in carbachol-mediated inhibition of isoproterenol-stimulated adenylyl cyclase.[60] These results point to a specific age-related decline in myocardial A_1-adenosine receptor function which may be related to the fact that myocardial muscarinic receptors are found in high density but have a low affinity for agonists, while the A_1-adenosine receptors have a high affinity for agonists, but are low in density.[5,95,96] Because of the different affinities for their respective natural ligands, the two receptor types may subserve different physiological functions. The low affinity cholinergic receptors allow for a rapid dissociation of the neurotransmitter and are thus suited for mediating stimuli that vary rapidly over time (i.e., fine temporal tuning). On the other hand, adenosine is expected to have a slower dissociation rate from its high affinity receptors, and thus the latter are better suited to provide tonic background modulation of function.[5] Since these two receptor types have different kinetic characteristics and physiological functions, it is not surprising that they do not share the same fate during aging.

A_1-ADENOSINE RECEPTOR LIGAND BINDING IN THE HEART

Radioligand binding studies were conducted to test whether A_1-adenosine receptor density and affinity change with age. Using the selective A_1-adenosine receptor ligand, [³H]8-cyclopentyl-1,3-dipropylxanthine ([³H]DPCPX), similar specific binding patterns were noted in ventricular membranes obtained from adult (6-month-old) and old (24-month-old) Fischer rats. There were no significant differences in receptor densities or equilibrium dissociation constants, between the two age groups (Fig. 5.1A). As described above, A_1-adenosine receptors belong to the G protein-coupled receptor superfamily, of which many appear to exist in two different agonist-affinity states, depending on whether they are coupled to their associated G proteins.[97] These receptors, when in their high affinity/G protein coupled state, may be the receptors

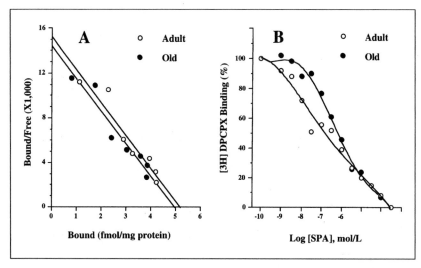

Fig. 5.1. Specific [³H]DPCPX binding in ventricular membranes of young (6-month-old) and old (24-month-old) Fischer 344 male rats. Representative Scatchard analysis of the binding data obtained, using 0.05-1.6 nmol/L [³H]DPCPX, shows similar linear regression curves for young and old rats (A). Typical agonist (SPA) displacement curves of [³H]DPCPX binding, performed in the presence of 0.3 nmol/L [³H]DPCPX with increasing concentrations of SPA (10^{-9}-$10^{-3.5}$ mol/L), were obtained from young and old rats (B).

which are ready to efficiently transduce signals to their downstream effectors. For the assessment of the effect of aging on the ratio of high/low affinity A_1-adenosine receptors, the agonist (SPA) displacement of [³H]DPCPX binding was carried out in myocardial ventricular membranes. In both adult and old animals, 5'-guanylylimidodiphosphate (Gpp(NH)p), a nonhydrolable GTP analog, was able to totally convert the high affinity portion of the A_1-adenosine receptors to their low affinity state, further demonstrating the G protein-coupled nature of the high affinity A_1-adenosine receptors.[61] While the total number of A_1-adenosine receptors was unaltered, the ratio of receptors in the high affinity state was reduced in tissue obtained from the aged animals (Fig. 5.1B), suggesting that in the ventricular myocardium the number of A_1-adenosine receptors which are coupled to G proteins is altered during aging.

DIRECT ASSESSMENT OF A_1-ADENOSINE RECEPTOR/G PROTEIN COUPLING

G proteins are trimeric molecules which consist of three subunits. To date, at least sixteen α subunits, five β subunits and seven γ subunits have been identified and theoretically, a large number of trimeric combinations of these subunits is possible. This diversity in the combinations of G protein subunits suggests that signal transduction via G protein-coupled receptors may be uniquely individualized and selective with regard to the information they transmit. The A_1-adenosine receptor is known to inhibit adenylyl cyclase via a PTX-sensitive G protein.[84] Reconstitution studies further documented that G_{i1}, G_{i2}, G_{i3} and G_o proteins may all be involved in the coupling of A_1-adenosine receptor to adenylyl cyclase.[21,85,98,99] Some G protein-coupled receptors may be linked to their G proteins in the absence of agonist and this interaction may be enhanced by receptor activation.[100-102] Employing an immuno-coprecipitation strategy,[103] we monitored the direct coupling of A_1-adenosine receptors to their G proteins. Although all the G proteins (G_s, G_{i1}, G_{i2}, G_{i3}, G_o and G_q) tested were present in ventricular membranes of Fischer rats, A_1-adenosine receptor binding sites, which were monitored by [^{3}H]DPCPX binding, were immunoprecipitated only with antisera raised against $G_{\alpha i3}$ and $G_{\alpha o}$ proteins.[61] In the absence of agonist, the coupling of A_1-adenosine receptor to G_{i3} and G_o proteins was reduced in old (24 month old) as compared to adult (6 month old) rats (Fig. 5.2). This result is well correlated with the agonist displacement experiment which indicated an age-associated reduction in ventricular high affinity A_1-adenosine receptor binding sites, thereby giving further support to the hypothesis of uncoupling of A_1-adenosine receptors from their associated G proteins in the aged myocardium. Moreover, agonist-enhanced coupling of A_1-adenosine receptors to G_{i3} and G_o proteins was also found to be diminished in ventricular membranes of old rats (Fig. 5.2). Thus, the age-associated impairment in A_1-adenosine receptor function (adenylyl cyclase inhibition) may be explained, at least in part, by the reduced coupling of these receptors with their G proteins.

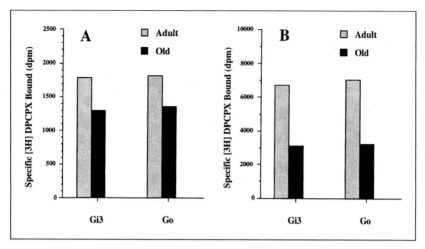

Fig. 5.2. Coupling of A_1 adenosine receptors with $G_{\alpha i3}$ and $G_{\alpha o}$ proteins in ventricular membranes of young (6-month-old) and old (24-month-old) Fischer 344 male rats. Ventricular membranes were solubilized and immunoprecipitated by incubation with antisera raised against $G_{\alpha i3}$ or $G_{\alpha o}$ proteins. The immunoprecipitates were resuspended and incubated with 1 nmol/L [^{3}H]DPCPX to determined A_1 adenosine receptors. A representative experiment showing basal coupling is shown in A. Agonist-enhanced coupling of A_1 adenosine receptors with $G_{\alpha i3}$ or $G_{\alpha o}$ proteins was determined by incubating membranes with 1 μmol/L SPA for 10 minutes prior to solubilization and immunoprecipitation. A representative experiment showing SPA-stimulated coupling is shown in B.

The mechanism for the age-related A_1-adenosine receptor/G protein uncoupling is presently not known. However, age-associated changes in phosphorylation of G protein coupled receptors, in GTP binding to membrane G proteins, in altered membrane constituents or membrane fluidity may contribute to this phenomenon. Since the level of adenosine in myocardial interstitial space may be higher in old vs. adult animals,[24-26] the reduction in A_1-adenosine receptor function, mediated by the uncoupling of receptors from G proteins, may be due to an adaptive response to this receptor overstimulation—a mechanism which is known to desensitize many other G protein-coupled receptors. Indeed, recent studies have demonstrated that long-term stimulation of A_1-adenosine receptors results in their desensitization.[104-106] On the other hand, the increase in production of adenosine in ventricles of old rats may be a compensatory response which develops during aging in an attempt to maintain the protective effects of adenosine as A_1-adenosine receptor function decreases during aging. Either way, there is an age-related decline in ventricular A_1-adenosine receptor function. Increased adenosine production during aging may compensate for the defect in A_1-adenosine receptor function and maintain the physiological function of adenosine. Under stress situations such as myocardial ischemia, the production of adenosine increases dramatically, but this increase may be much less extensive in aged animals.[34] Thus, the uncoupling of A_1-adenosine receptors from the G proteins could leave the myocardium critically unprotected during stress and may be part of the reason for the increased susceptibility of the aged myocardium to ischemic injury.

SUMMARY

Adenosine regulates cardiac function in a number of different ways. It affects cardiac electrophysiology, myocardial contractility and coronary blood flow. These actions in the heart are age-dependent, and at least in part, contribute to the known age-related changes in cardiovascular function. However, since there are so many conflicting results which may be related to differences in animal species and specific experimental conditions used in different studies, the results should be interpreted with great caution. Furthermore, aging is frequently associated with increased pathology and disease state which may further complicate cardiovascular function. In future investigations, the putative role of adenosine in the pathogenesis of disorders or in the susceptibility to pathogenic factors of the aged heart need to be determined.

ACKNOWLEDGMENT

The original research reported in this chapter was in part supported by grants AG 07700 from the National Institute on Aging, and T32-AG 00131 NIH.

REFERENCES

1. Drury AN, Szent-Gyorgyi A. The physiological activity of adenosine compounds with especial reference to their action upon the mammalian heart. J Physiol (Lond) 1929; 68:213-237.

2. Berne RM. Cardiac nucleotides in hypoxia: Possible role in regulation of coronary blood flow. Am J Physiol 1963; 204:317-322.

3. Gerlach E, Deuticke B, Dreisbach RH. Der Nucleotide-Abbau im Herzmuskel bei Sauerstoffmangel und seine mogliche Bedeutung fur die Coronardurchblutung. Naturwissenschaften 1963; 50:228-229.

4. Belardinelli L, Linden J, Berne RM. The cardiac effects of adenosine. Prog Cardiovas Dis 1989; 32:73-97.

5. Olsson RA, Pearson JD. Cardiovascular purioceptors. Physiol Rev 1990; 70:761-809.

6. Pelleg A, Belardinelli L. Cardiac electrophysiology and pharmacology of adenosine: Basic and clinical aspects. Cardiovas Res 1993; 27:54-61.

7. Mubagwa K, Mullane K, Flameng W. Role of adenosine in the heart and circulation. Cardiovas Res 1996; 32:797-813.

8. Pelleg A. Cardiac electrophysiology and pharmacology of adenosine and ATP: Modulation by the autonomic nervous system. J Clin Pharmacol 1987; 27:366-372.

9. Schrader J, Baumann G, Gerlach E. Adenosine as inhibitor of myocardial effects of catecholamine. Pflugers Arch 1977; 372:29-35.

10. Belardinelli L, Vogel S, Linden J et al. Antiadrenergic action of adenosine on ventricular myocardium in embryonic chick hearts. J Mol Cell Cardiol 1982; 14:291-294.

11. Dobson JG Jr. Mechanism of adenosine inhibition of catecholamine-induced responses in heart. Circ Res 1983; 52:151-160.

12. Samet MK, Rutledge CO. Antagonism of the positive chronotropic effect of norepinephrine by purine nucleosides in rat atria. J Pharmacol Exp Ther 1984; 232:106-110.

13. Fredholm BB, Dunwiddie TV. How does adenosine inhibit transmitter release? Trends Pharmacol Sci 1988; 9:133-134.

14. Burnstock G. Purinergic receptors. J Theor Biol 1976; 62:491-593.

15. Burnstock G. Purinergic receptors in the heart. Circ Res 1980; 46:176-186.

16. Von Calker D, Muller DM, Hamprecht B. Adenosine regulates via two different types of receptors, the accumulation of cyclic AMP in cultured brain cells. J Neurochem 1979; 33:999-1005.

17. Londos C, Cooper DMF, Wolff J. Subclasses of external adenosine receptors. Proc Natl Acad Sci (USA) 1980; 77:2551-2554.

18. Jacobson KA, De LaCruz R, Schulick R et al. 8-substitute xanthines as antagonists at A_1- and A_2-adenosine receptors. Biochem Pharmacol 1988; 37:3653-3661.

19. Tucker AL, Linden J. Cloned receptors and cardiovascular responses to adenosine. Cardiovas Res 1993; 27:62-67.

20. Olah ME, Stiles GL. Adenosine receptor subtypes: Characterization and therapeutic regulation. Annu Rev Pharmacol Toxicol 1995; 35:581-606.

21. Stiles GL. Adenosine receptors. J Biol Chem 1992; 267:6451-6454.

22. Lerman BB, Belardinelli L. Cardiac electrophysiology of adenosine. Basic and clinical concepts. Circulation 1991; 83:1499-1509.
23. Docherty JR. Cardiovascular responses in aging: a review. Pharmacol Rev 1990; 42:103-125.
24. Dobson JG Jr, Fenton RA, Romano FD. Increased myocardial adenosine production and reduction of β-adrenergic contractile response in aged hearts. Circ Res 1990; 66:1391-1390.
25. Dobson JG Jr, Fenton RA. Adenosine inhibition of β-adrenergic induced responses in aged hearts. Am J Physiol 1993; 265:H494-H503.
26. Headrick JP. Impact of aging on adenosine levels, A_1/A_2 responses, arrhythmogenesis, and energy metabolism in rat heart. Am J Physiol 1996; 270:H897-H906.
27. Hoffman BB, Chang H, Farahbakhsh Z et al. Inhibition of lipolysis by adenosine is potentiated with age. J Clin Inves 1984; 74:1750-1755.
28. Ethier MF, Hickler RB, Dobson JG Jr, Aging increases adenosine and inosine release by human fibroblast cultures. Mech Ageing Develop 1989; 50:159-168.
29. Matherne GP, Headrick JP, Coleman SD et al. Interstitial transudate purines in normoxic and hypoxic immature and mature rabbit hearts. Pediatr Res 1990; 28:348-353.
30. Murphy E, Holland MJC, Cox RP. Adenosine deaminase activity in human diploid skin fibroblasts varies with the age of the donor. J Med 1978; 9:237-244.
31. Jamal Z, Saggerson ED. Enzymes involved in adenosine metabolism in rat white and brown adipocytes. Biochem J 1987; 49:519-531.
32. Scholar EM, Rashidian M, Heidrick ML. Adenosine deaminase and purine nucleoside phospholyase activity in spleen cells of aged mice. Mech Ageing Develop 1980; 12:323-329.
33. Crosti F, Ciboddo GF, Barbier MC et al. Evidence for adenosine deaminase lymphocyte system impairment in ageing. Boll Ist Sieroter Milan 1987; 66:282-288.
34. Ramani K, Lust WD, Whittingham et al. ATP and catabolism and adenosine generation during ischemia in the aging heart. Mech Ageing Develop 1996; 89:113-124.
35. Funaya H, Kitakaze M, Node K et al. Plasma adenosine levels increase in patients with chronic heart failure. Circulation 1997; 95:1363-1353.
36. Belardinelli L. Adenosine system in the heart. Drug Dev Res 1993; 28:263-267.
37. Belardinelli L, Shryock JC, Song Y et al. Ionic basis of the electrophysiological actions of adenosine on cardiomyocytes. FASEB J 1995; 9:359-365.
38. Martynyuk AE, Kane KA, Cobbe SM et al. Adenosine increases potassium conductance in isolated rabbit atrioventricular nodal myocytes. Cardiovas Res 1995; 30:668-675.

39. Young ML, Ramza BM, Tan RC et al. Adenosine and hypoxia effects on atrioventricular node of adult and neonatal rabbit hearts. Am J Physiol 1987; 253:H192-H198.

40. Hatae J, Sperelakis N, Wahler GM. Development of the response to adenosine during organ culture of young embryonic chick hearts. J Dev Physiol 1989; 11:142-145.

41. Mudumbi RV, Olson RD, Hubler RE et al. Age-related effects in rabbit hearts of N^6-R-phenylisopropyladenosine, an adenosine A_1 receptor agonist. J Geront 1995; 50A:B351-B357.

42. Montamat SC, Olson RD, Mudumbi RV et al. Age-related characterization of atrial adenosine A_1 receptor activation: Direct effects on chronotropic and inotropic function in the Fischer 344 rat. J Geront 1996; 51A: B239-B246.

43. DiGennaro M, Bernabei R, Sgadari A et al. Age-related differences in isolated rat sinus node function. Basic Res Cardiol 1987; 82:530-536.

44. Xu J, Gao E, Friedman E et al. Effects of aging on the negative chronotropic and anti-β adrenergic actions of adenosine in the rat heart. Personal communication.

45. Watt AH, Bayer A, Routledge PA et al. Adenosine-induced respiratory and heart rate changes in young and elderly adults. Br J Clin Pharmacol 1989; 27:265-267.

46. Suteparuk S, Nies AS, Andros E et al. The role of adenosine in promoting cardiac beta-adrenergic subsensitivity in aging human. J Geront 1995; 50A:B128-B134.

47. Koumi SI, Wasserstrom A. Acetylcholine-sensitive muscarinic K^+ channels in mammalian ventricular myocytes. Am J Physiol 1994; 266:H1812-H1821.

48. Ito H, Hosoya Y, Inanobe A et al. Acetylcholine and adenosine activate the G protein-gated muscarinic K^+ channel in ferret ventricular myocytes. Nauuyn-Schmiedebergs Arch Pharmacol 1995; 352:610-617.

49. Qu Y, Himmel HM, Campbell DL et al. Modulation of basal L-type Ca^{2+} current by adenosine in ferret isolated right ventricular myocytes. J Physiol (Lond) 1993; 471:269-293.

50. Bohm M, Pieske B, Ungerer M et al. Characterization of A_1 adenosine receptors in atrial and ventricular myocardium from diseased human hearts. Circ Res 1989; 65:1201-1211.

51. Dobson JG Jr. Reduction by adenosine of the isoproterenol induced increase in cyclic adenosine 3',5'-monophosphate formation and glycogen phosphorylase activity in rat heart muscle. Circ Res 1978; 43:785-792.

52. Shen WK, Kurachi Y. Mechanisms of adenosine-mediated actions on cellular and clinical cardiac electrophysiology. Mayo Clin Proc 1995; 70:274-291.

53. Romano FD, Fenton RA, Dobson JG Jr. The adenosine Ri agonist, phenylisopropyladenosine, reduces high affinity isoproterenol binding to the β-adrenergic receptor of rat myocardial membranes. Second Messengers Phosphoprotein 1988; 12:29-43.

54. Cothran DL, Lloyd TR, Linden J et al. Ontogeny of rat myocardial A_1 adenosine receptors. Biol Neonate 1995; 68:111-118.

55. Matherne GP, Headrick JP, Berne RM. Ontogeny of adenosine response in guinea pig heart and aorta. Am J Physiol 1990; 259: H637-H642.

56. Kumar R, Akita T, Joyner RW. Adenosine and carbachol are not equivalent in their effects on L-type calcium current in rabbit ventricular cells. J Mol Cell Cardiol 1996; 28:403-415.

57. Thakkar JK, Sperelakis N. Changes in cyclic nucleotide levels during embryonic development of chick hearts. J Dev Physiol 1987; 9:497-505.

58. Sawmiller DR, Fenton RA, Dobson JG Jr. Myocardial adenosine A_1 and A_2 receptor activities during juvenile and adult stages of development. Am J Physiol 1996; 271:H235-H243.

59. Romano FD, Dobson JG Jr. Adenosine attenuation of isoproterenol-stimulated adenylyl cyclase activity is enhanced with aging in the adult heart. Life Sci 1996; 58:493-502.

60. Gao E, Snyder DL, Johnson MD et al. The effect of age on adenosine A_1 receptor function in the rat heart. J Mol Cell Cardiol 1997; 29:593-602.

61. Cai G, Wang HY, Gao E et al. Reduced adenosine A_1 receptor-G protein coupling in rat ventricular myocardium during aging. Cir Res 1997; 81:1065-1071.

62. Vials A, Burnstock G. A_2-purinoceptor-mediated relaxation in the guinea-pig coronary vasculature: a role for nitric oxide. Br J Pharmacol 1993; 109:424-429.

63. Zanzinger J, Bassenge E. Coronary vasodilation to acetylcholine, adenosine and bradykinin in dogs: effects of inhibition of NO-synthesis and captopril. Eur Heart J 1993; 14(Suppl I):164-168.

64. Daut J, Marier-Rudolph W, von Beckerath N et al. Hypoxic dilation of coronary arteries is mediated by ATP-sensitive potassium channels. Science 1990; 247:1341-1344.

65. Belloni FL, Hintze TH. Glibenclamide attenuates adenosine-induced bradycardia and coronary vasodilation. Am J Physiol 1991; 261: H720-H727.

66. Furukawa S, Satoh K, Taira N. Opening of ATP-sensitive K^+ channels responsible for adenosine A_2 receptor-mediated vasodepression does not involve a pertussis toxin-sensitive G protein. Eur J Pharmacol 1993; 236:255-262.

67. Nakhostine N, Lamontagne D. Adenosine contributes to hypoxia-induced vasodilation through ATP-sensitive K^+ channel activation. Am J Physiol 1993; 265:H1289-H1293.

68. Fozard JR, Carruthers AM. The cardiovascular effects of selective adenosine A_1 and A_2 receptor agonists in the pithed rat: No role for glibenclamide-sensitive potassium channels. Naunyn-Schmiedebergs Arch Pharmacol 1993; 347:192-196.

69. Niiya K, Uchida S, Tsuji T et al. Glibenclamide reduces the coronary vasoactivity of adenosine receptor agonists. J Pharmacol Exper Ther 1994; 271:14-19.

70. Shimizu I, Toda N. Alterations with age of the response to vasodilator agents in isolated mesenteric arteries of the beagle. Br J Pharmacol 1986; 89:769-778.

71. Cook JJ, Wailgum TD, Vasthare US et al. Age-related alterations in the arterial microvasculature of skeletal muscle. J Geront 1992; 47A:B83-B88.

72. Buss DD, Hannemann WW 3rd, Posner P. Maturation of coronary responsiveness to exogenous adenosine in the rabbit. Basic Res Cardiol 1987; 82:290-296.

73. Laudignon N, Ardanda JV, Varma DR. Effects of adenosine and its analogues on isolated internal carotid arteries from newborn and adult pigs. Biol Neonate 1990; 58:91-97.

74. Toda N, Bian K, Inoue S. Age-related changes in the response to vasoconstrictor and dilator agents in isolated beagle coronary arteries. Naunyn-Schmiedebergs Arch Pharmacol 1987; 336:359-364.

75. Wakade AR, Wakade TD. Inhibition of noradrenaline release by adenosine. J Physiol (Lond) 1978; 282:35-49.

76. Hedqvist P, Fredholm BB. Inhibitory effect of adenosine on adrenergic neuroeffector transmission in the rabbit heart. Acta Physiol Scand 1979; 105:120-122.

77. Richardt G, Waas W, Kranzhofer R et al. Adenosine inhibits exocytolic release of endogenous noradrenaline in rat heart: A protective mechanism in early myocardial ischemia. Circ Res 1987; 61:117-123.

78. Martens D, Lohse M, Rauch B. Pharmacological characterization of A_1 adenosine receptors in isolated rat ventricular myocytes. Naunyn-Schmiedebergs Arch Pharmacol 1987; 336:342-348.

79. Henrich M, Peper HM, Schrader J. Evidence for adenylyl cyclase-coupled A_1 adenosine receptors on ventricular cardiomyocytes from adult rat and dog heart. Life Sci 1987; 41:2381-2388.

80. Wennmalm M, Fredholm BB, Hedqvist P. Adenosine as a modulator of sympathetic nerve-stimulation-induced release of noradrenaline from the isolated rabbit heart. Acta Physiol Scand 1988; 132:487-494.

81. Richardt G, Waas W, Kranzhofer R et al. Interaction between the release of adenosine and noradrenaline during sympathetic stimulation: a feed-back mechanism in rat heart. J Mol Cell Cardiol 1989; 21:269-277.

82. Schutz W, Stroher M, Freissmuth M et al. Adenosine receptors mediate a pertussis toxin-insensitive prejunctional inhibition of noradrenaline release on a papillary muscle model. Naunyn-Schmiedebergs Arch Pharmacol 1991; 343:311-316.

83. Snyder DL, Wang W, Pelleg A et al. The effect of age on A_1-adenosine receptor mediated inhibition of norepinephrine release in the rat heart. Personal communication.

84. Olah ME, Stiles GL. Adenosine receptors. Annu Rev Physiol 1992; 54:211-225.

85. Munshi R, Pang IH, Sternweis et al. A_1 adenosine receptor of bovine brain couple to guanine nucleotide-binding proteins G_{i1}, G_{i2}, and G_o. J Biol Chem 1991; 266:22285-22289.

86. Jocker R, Linder ME, Hohenegger M et al. Species difference in the G protein selectivity of the human and bovine A_1-adenosine receptor. J Biol Chem 1994; 269:32077-32084.

87. Linden J, Hollen CE, Patel A. The mechanism by which adenosine and cholinergic agents reduce contractility in rat myocardium: Correlation with cyclic adenosine monophosphate and receptor densities. Circ Res 1985; 56:728-735.

88. LaMonica DA, Frohloff N, Dobson JG Jr. Adenosine inhibition of catecholamine-stimulated cardiac membrane adenylate cyclase. Am J Physiol 1985; 248:H737-H744.

89. West GA, Isenberg G, Belardinelli L. Antagonism of forskolin effects by adenosine in isolated hearts and ventricular myocytes. Am J Physiol 1986; 250:H769-H777.

90. Ma H, Yu HJ, Green RD. Adenosine mediated inhibition of cardiac adenylyl cyclase activity may involve multiple receptor subtypes. Naunyn-Schmidebergs Arch Pharmacol 1994; 349:81-86.

91. Bruns RF, Lu GH, Pugsley TA. Characterization of the A_2 adenosine receptor labeled by [³H]NECA in rat striatal membranes. Mol Pharmacol 1986; 29:331-346.

92. Jacobson KA, Nikodijevic O, Ji XD et al. Synthesis and biological activity of N^6-(p-sulfophenyl)alkyl and N^6-sulfoakyl derivatives of adenosine: water soluble and peripherally selective adenosine agonists. J Med Chem 1992; 35:4143-4149.

93. Kemmer M, Jakob H, Nawrath H. Pronounced cholinergic but only moderate purinergic effects in isolated atrial and ventricular heart muscle from cats. Br J Pharmacol 1989; 97:1191-1198.

94. Bohm M, Gierschik P, Jakobs KH et al. Increase in $G_{\alpha i}$ in human hearts with dilated but not ischemic cardiomyopathy. Circulation 1990; 82:1249-1265.

95. Endoh M, Kushida H, Norota I et al. Pharmacological characteristics of adenosine-induced inhibition of dog ventricular contractility: dependence on the preexisting level of β adrenoceptor activation. Naunyn-Schmiedebergs Arch Pharmacol 1991; 344:70-78.

96. Bohm M, Gierschik P, Schwinger RHG et al. Coupling of M-cholinoceptors and A_1 adenosine receptors in human myocardium. Am J Physiol 1994; 266:H1951-H1958.

97. Gilman AG. G proteins: Transducers of receptor-generated signals. Annu Rev Biochem 1987; 56:615-659.

98. Freissmuth M, Schutz W, Linder ME. Interactions of the bovine brain A_1-adenosine receptor with recombinant G protein alpha-subunits. Selectivity for rG_i alpha-3. J Biol Chem 1991; 266:17778-17783.

99. Asano T, Shinohara H, Morishita R et al. The G protein G_o in mammalian cardiac muscle: localization and coupling to A_1 adenosine receptors. J Biochem 1995; 117:183-189.

100. Tian WN, Duzic E, Lanier SM et al. Determinants of α_2-adrenergic receptor activation of G proteins: Evidence for a precoupled receptor/G protein state. Mol Pharmacol 1994; 45:524-531.

101. Shi AG, Deth RC. Precoupling of alpha-2B adrenergic receptors and G proteins in transfected PC-12 cell membrane: Influence of pertussis toxin and a lysine-directed cross-linker. J Pharmacol Experi Ther 1994; 271:1520-1527.

102. Gurdal H, Friedman E, Johnson MD. β-Adrenoceptor-$G_{\alpha s}$ coupling decreases with age in rat aorta. Mol Pharmacol 1995; 47:772-778.

103. Friedman E, Butkerait P, Wang HY. Analysis of receptor-stimulated and basal guanine nucleotide binding to membrane G protein by sodium dodecyl sulfate-polyacrylamide gel electrophoresis. Anal Biochem 1993; 214:171-178.

104. Ramkumar V, Olah ME, Jacobson KA et al. Distinct pathways of desensitization of A_1- and A_2-adenosine receptors in DD1 MF-2 cells. Mol Pharmacol 1991; 40:639-647.

105. Palmer TM, Benovic JL, Stiles GL. Molecular basis for subtype-specific desensitization of inhibitory adenosine receptors. J Biol Chem 1996; 271:15272-15278.

106. Bhattacharya S, Linden J. Effects of long-term treatment with the allosteric enhancer, PD81,723, on Chinese hamster ovary cells expressing recombinant human A_1 adenosine receptors. Mol Pharmacol 1996; 50:104-111.

CHAPTER 6

Direct and Indirect Effects of Extracellular ATP on Cardiac Myocytes

Amir Pelleg and Guy Vassort

INTRODUCTION

Adenosine 5'-triphosphate (ATP) plays a major role in intracellular metabolism and energetics. Since the first observations by Drury and Szent-Gyorgyi[1] of the pronounced cardiovascular effects of extracellular ATP, numerous studies have documented its various effects in different cells, tissues and organs, mediated by specific cell-surface receptors (P_2-purinoceptors) distinct from adenosine receptors (i.e., P_1-purinoceptors).[2-4] Thus, ATP can be viewed as a local regulator which could play an important role under physiologic and pathophysiologic conditions.

There are several sources for extracellular ATP: ATP is stored in relatively large amounts in platelets and is released during platelet activation.[5-7] Similarly, ATP is stored in red blood cells from which it is released under conditions of imbalance between O_2 supply and O_2 demand.[8-10] In addition, several biological substances as

Effects of Extracellular Adenosine and ATP on Cardiomyocytes,
edited by Amir Pelleg and Luiz Belardinelli. © 1998 R.G. Landes Company.

well as increased flow can induce the release of ATP from vascular endothelial cells[11-15] and smooth muscle cells.[16,17] ATP is also released from neural elements as a cotransmitter[18] and from exercising skeletal muscle.[19]

In the heart, ATP is released into the extracellular fluid under various conditions. Specifically, ATP release is evoked by sympathetic nerve stimulation and by catecholamines.[20-23] In addition, ATP is released in the heart during acute myocardial ischemia[24] and from cardiac myocytes in response to hypoxia.[25,26]

The mechanism by which ATP is transported across the cell membrane is not fully known. In recent years the existence of adenine nucleotides binding cassette (ABC) family of proteins has been proposed.[27] These proteins were suggested to be a regulatory component of an ion-channel-regulator complex, such as the cystic fibrosis transmembrane conductance regulator (CFTR) which acts as an ATP channel enabling intracellular ATP to cross the cell membrane and stimulate cell surface receptors.[27] However, whether the CFTR channel is permeable to ATP has been the subject of some controversy.[28-30]

Extracellular ATP induces multiple functional changes in cardiac cells. These effects can be due to either direct action of the nucleotide on cardiac cells or indirect action mediated by the parasympathetic limb of the autonomic nervous system. In the latter case, the neurotransmitter acetylcholine is the active substance at the myocardial cell surface. The text below details both the direct and the indirect actions of ATP, their basic and clinical implications as well as suggested future research directions.

DIRECT ACTIONS OF ATP ON CARDIAC CELLS

ATP Modulates Transmembrane Ionic Currents: K⁺ Currents

A number of potassium channels are present in cardiac myocytes that determine the shape of the action potential and frequency of beating. Several observations have indicated that most of these channels are regulated by extracellular ATP. In 1988 Friel and Bean[31] reported that extracellular ATP activates two different ionic conductances in bullfrog atrial cells; one transient, probably Na^+ and other cations, and the other sustained, probably K^+, manifesting linear and inwardly current-voltage relation, respectively. Adenosine, AMP, ADP, ITP and UTP were completely ineffective in activating either conductance.[31] In a subsequent study, these authors have found that extracellular ATP but not adenosine, activates an inwardly rectifying potassium channel in calf atrial myocytes.[32] This channel seemed nearly identical to the potassium channel activated by acetylcholine in these cells; the conductance of the channel activated by ATP and acetylcholine was 30 and 31 pS, respectively.[32] In view of the similarity between the electrophysiologic actions of adenosine and acetylcholine in the mammalian heart, it is interesting that adenosine failed to induce potassium conductance in bovine atrial myocytes. Indeed, in a more recent study using one-and two-day-old rat atrial myocytes, extracellular ATP and adenosine activated kinetically similar potassium channels (i.e., single channel conductance and mean open time of 32.0 ± 0.2 pS and 0.5 ± 0.1 msec, respectively, vs. 31.3 ± 0.3 pS and 0.9 ± 0.1 msec, respectively).[33] The muscarinic cholinergic receptor and the A_1-adenosine receptor are known to be directly coupled to a K^+ channel ($K^+_{Ach,Ado}$ channel) via a pertussis toxin (PTX) sensitive G_K protein. In guinea pig atrial cells extracellular ATP shortens the action potential; this effect is mediated by a P_2-purinoceptor directly coupled to $K^+_{Ach,Ado}$ channel through a PTX-sensitive G_K protein, analogous to the activation of the channel by either acetylcholine or adenosine.[34] In another recent study using isolated guinea pig atrial myocytes, extracellular ATP (10 µM) transiently activated $I_{K,Ach,Ado}$; however, when this current was preactivated with either carbochol or adenosine, ATP produced a transient increase followed by a sustained decrease of the current.[35] The inhibition of I_k, Ach by extracellar ATP probably occurs following the activation of a PTX-insensitive G protein and a cytosolic second messenger but does not involve protein kinase C.[35a] These data were interpreted[35] as a possible explanation for the biphasic inotropic effect, i.e., rapid negative followed by slow positive inotropic effect, of extracellular ATP in rat atrial preparation.[36] Similar increase in the $I_{K,Ach,Ado}$ is elicited by extracellular ATP in rat ventricular myocytes (Aimond, Lorente and Vassort, unpublished observation).

It was also reported that ATP activated the delayed rectifier K[+] current (I_K), which is slowly activated during the action potential plateau and facilitates repolarization, and whose deactivation contributes to depolarization of pacemaker cells.[37] More recently the same group reported that only the slow component of the delayed rectifier K[+] current was selectively enhanced by ATP[37a].

Another type of K[+] channel, the intracellular ATP-sensitive channel (K^+_{ATP} channel), which is inhibited by intracellular ATP,[38] can also be regulated by extracellular ATP. K^+_{ATP} channel activation during acute ischemia/hypoxia has been shown to exert a protective effect on the heart.[39,40] Studies in cardiac myocytes in vitro have suggested that the activation of A_1-adenosine receptors could result in the activation of K^+_{ATP} channels.[41,42] However, at least in the hypoxic guinea pig heart in vivo, endogenous adenosine failed to activate K^+_{ATP} channels.[43] The K^+_{ATP} channel consists of a weak inward rectifier subunit Kir 6.2, plus a member of the adenine nucleotide binding cassette (ABC) superfamily, SUR2.[44] Recent studies have shown that extracellular ATP enhances the current flow through this channel ($I_{K,ATP}$) once it has been partially activated under conditions of metabolic stress (i.e., 100 μM of intracellular ATP);[45] the enhancement of $I_{K,ATP}$ by extracellular ATP was inhibited by cholera toxin as well as by inhibition of adenylyl cyclase.[46] Thus, it has been suggested that the mechanism of this effect is the G_s dependent activation of adenylyl cyclase which causes increased cAMP production and thereby reduced local levels of intracellular ATP.[46] Several analogs of ATP, i.e., α,β-mATP, 2mSATP and ATPγS exerted a similar effect to that of ATP, while UTP and ADP had a relatively small effect and AMP and adenosine had no effect.[45,46]

ATP MODULATES TRANSMEMBRANE IONIC CURRENTS: Ca^{2+} CURRENTS

Two decades ago Goto et al,[47] had found that extracellular ATP and ADP enhanced calcium inward current (I_{Ca}) and I_{Ca}-dependent phasic tension in muscle bundles isolated from the right atrium of the bullfrog. In a subsequent study, the same group determined that the action of ATP on I_{Ca} and tension did not require the hydrolysis of ATP and was probably mediated by a receptor located at the outer surface of the affected cell membrane.[48]

De Young and Scarpa[49] used a fluorescent Ca^{2+} indicator to show that extracellular ATP induces Ca^{2+} transients in isolated rat cardiac ventricular myocytes. This action of ATP was not mimicked by either the ATP metabolites ADP, AMP and adenosine or by other nucleotide triphosphates such as GTP, ITP, UTP, CTP and TTP.[49] Pretreatment of the cells with 0.2-1 µM norepinephrine potentiated the action of ATP on Ca^{2+} transients; the effect of norepinephrine was mediated by β_1-adrenoceptors activation.[49] This study confirmed earlier preliminary data by Sharma and Sheu.[50] It was later determined that the increase in $[Ca^{2+}]_i$ induced by ATP was due to augmentation of Ca^{2+} influx[51] which was associated with increased amplitude of contraction.[52]

The potentiation and attenuation of ATP-induced Ca^{2+} transients by BayK8644 and nifedipine and verapamil, respectively, confirmed that Ca^{2+} influx induced by ATP involves voltage-sensitive Ca^{2+} channels.[53-55] In addition, depletion of intracellular Ca^{2+} store diminished the effect of ATP indicating that Ca^{2+} dependent Ca^{2+} release from the sarcoplasmic reticulum is mechanistically involved in the increase of $[Ca^{2+}]_i$ induced by extracellular ATP.[53,54]

In single cells isolated from frog ventricle, ATP (1 µM) increased I_{Ca} by up to two-fold; at higher ATP concentrations the increase in I_{Ca} was smaller and at 100 µM, ATP reduced this current.[56] The ATP-induced increase in Ca^{2+} current was prevented by perturbations which block either signal transduction pathway to phospholipase C (PLC) or its activity.[56] These data were interpreted to suggest that ATP-induced increase in Ca^{2+} current in frog ventricular myocytes is mediated by P_2-purinoceptor and phosphoinositide turnover.[56] ATP (10 μM) increased also the L-type Ca^{2+} current in single cells isolated from rat ventricular myocardium;[55,57-61] adenosine 5'-0-(3-thio triphosphate) (ATPγS) exerted a similar effect, but adenosine was much less effective and GTP, UTP, CTP and ITP were without effect.[57] Incubation of the cells with cholera toxin (10 μg/ml) for four hours resulted in significantly greater Ca^{2+} current density which could not be modified by ATP.[58] Thus, it seems that the activation of P_2-purinoceptor leads to an increase of Ca^{2+} current via the activation of cholera toxin-sensitive G protein; the latter was identified as an isoform of G_s.[61] Similarly, ATP increases the transient, high threshold Ca current (I_{CaT}) in frog atrial cells via a pathway that does not involve phosphorylation.[62]

Extracellular ATP-induced Ca^{2+} transients in ventricular myocytes are followed by a slower but larger $[Ca^{2+}]_i$ increase when extracellular P_i is increased.[63] This second phase of $[Ca^{2+}]_i$ response, induced also by ATPγS, was explained by the stimulation of Na^+-P_i cotransport by ATP which led to the stimulation of Na^+-Ca^{2+} exchange.[63] Zheng et al[64] and Christie et al,[55] have proposed the following mechanism for ATP-induced increase in $[Ca^{2+}]_i$ and its potentiation by β-adrenoceptor activation: ATP stimulates P_2-purinoceptors leading to an inward current (I_{ATP}) via activated nonselective cation channels and thereby the depolarization of the cell membrane. The latter activates the voltage dependent L-type Ca^{2+} channel, the number of which is increased by β-adrenoceptor activation, stimulation of adenylyl cyclase and cAMP dependent activation of protein kinase A (PKA) which phosphorylates L-type Ca^{2+} channels. And finally, Ca^{2+} induced Ca^{2+} release from the sarcoplasmic reticulum (SR).[64] Indeed, pretreatment of the cells with caffeine and ryanodine partially inhibited the $[Ca^{2+}]_i$ increase[55] and the activation of protein kinase C (PKC) reduced cAMP level, attenuated ATP-induced Ca^{2+} transients and its potentiation by norepinephrine without affecting the nonselective cation current induced by ATP.[65] More recently Zheng et al[66] have suggested that the induction of the I_{ATP} and I_{Ca} by extracellular ATP is mediated by two different mechanisms, i.e., the activation of a ligand binding channel and a signal transduction pathway involving a G protein, respectively. Both mechanisms seemed to be independent of either cAMP or phosphoinositidle turnover.[66]

In contrast to its actions in frog and rat ventricular myocytes and guinea pig and rabbit atrial myocytes,[67] extracellular ATP inhibited I_{Ca} in a time- and concentration-manner dependent in isolated ferret ventricular myocytes.[68] This action of ATP was independent of adenosine A_1 receptors but involved ATP binding to P_{2Y} purinoceptor and the activation of pertussis toxin (PTX) insensitive G protein.[69] It should be noted here that the ferret is unique also with regard to the electrophysiologic actions of adenosine. While adenosine has no direct effect on the ventricular myocardium of all mammalian species studied, it attenuates L-type Ca^{2+} current[70] and activates $I_{KAdo,Ach}$ in ferret ventricular myocytes.[71]

In rat ventricular myocytes, ATP analogs showed the following rank order of potency in increasing $[Ca^{2+}]_i$: ATP≥2mSATP> >ATPγS>d,βmATP≈β,γmATP≈β, imido-ATP.[54] The rank order of potency of ATP analogs in increasing I_{Ca} was 2mSATP≈ATPγS; in contrast, α,βmATP, β,γmATP had no significant effect.[60] The rank order of potency of ATP analogs in activating I_{ATP} was 2mSATP ATP, while α,βmATP, β,γmATP and β,γ-imido-ATP did not activate I_{ATP}.[60] These data were interpreted to suggest that the mediation by P_{2Y}-purinoceptors is mechanistically involved in the modulation by extracellular ATP of $[Ca^{2+}]_i$.[54,60]

In rat papillary and right ventricular muscles[72] and cultured ventricular myocytes of fetal mice,[73] extracellular ATP stimulated phosphoinositide hydrolysis and in the latter preparation also inhibited the isoproterenol-induced accumulation of cAMP. It was suggested that PLC and PTX sensitive G protein mediate these actions of ATP.[73] Based on the aforementioned studies it can be concluded that extracellular ATP can directly affect ionic permeabilities in cardiac ventricular myocytes. The action of ATP is characterized by species variability. At least in the mouse and rat, ATP increases $[Ca^{2+}]_i$ probably by stimulating cell surface P_2-purinoceptors

Table 6.1. Inotropic effect of extracellular ATP

Species	Preparation	Inotropic Effect	Reference
Frog	Left atrial bundle	+	47
			48
Frog	Atria	+/–	115
	Ventricular strip	+	115
Guinea pig	Left atria	–	116
Left atria		+	117
Rat	Left auricle	–	118
	Auricle strip*	+/–	57
	Atria	–	119
	Papillary muscle	+	72
	Ventricular strip	–	120
	Perfused heart	–	121,122
Cat	Atria	–	123

* Negative inotropic effect in control, positive inotropic effect in PTX-treated tissue.

and activating a nonselective cationic channel leading to depolarization of the cell membrane and Ca^{2+} dependent Ca^{2+} release from the SR. The exact signal transduction pathway mediating ATP's action is not conclusively determined. Table 6.1 summarizes the inotropic effects of ATP in different experimental preparations. Figure 6.1 is a schematic outline of the signal transduction pathways of ATP which affect $[Ca^{2+}]_i$.

In addition to its effects on I_K and I_{Ca}, extracellular ATP can affect inward Na^+ current (I_{Na}). Using the whole-cell patch clamp in rat single cardiac ventricular myocytes, Scamps and Vassort[74] have found that extracellular ATP in micromolar range, caused a leftward shift in both activation and availability characteristics of I_{Na} as was already shown for the L-type Ca current.[61] At hyperpolarized potentials, I_{Na} was increased due to the shift in activation, whereas at depolarized potentials, I_{Na} was decreased because of reduced availability.[74] ATPγS and α,βmATP exerted similar effects but UTP, β,γmATP, ADP and adenosine were without effect. The shifts

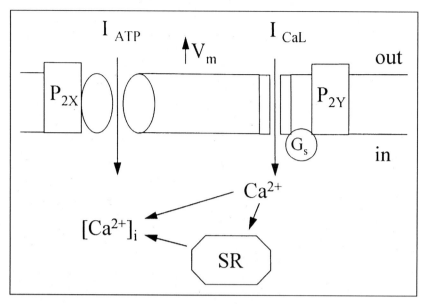

Fig. 6.1. Schematic outline of signal transduction pathways which mediate the action of extracellular ATP on intracellular Ca^{2+} level (Ca_i^{2+}). On the right, the stimulation of P_{2Y}-purinoceptor coupled to G_s protein leads to the activation of L-type Ca^{2+}-channel and the inward Ca^{2+} current (I_{CaL}) and causes Ca^{2+} release from the sarcoplasmic reticulum (Ca^{2+} dependent - Ca^{2+} release). This is in addition to inositol triphosphate (IP_3)-induced release (depicted also in Fig. 6.3). On the left, extracellular ATP can bind to P_{2X}-purinoceptor which activates the nonselective cationic channel. The resulting inward current (I_{ATP}) depolarizes the cell membrane potential (V_m) which leads to activation of the L-type Ca^{2+} channel (i.e., voltage dependent activation).

observed upon application of extracellular ATP were not affected by cholera toxin treatment suggesting that G_s protein and cAMP are not involved in this phenomenon.[74] This cell membrane potential dependent differential effect of ATP could be arrhythmogenic under pathophysiologic conditions associated with heterogenous resting cell membrane potential.

ATP MODULATES TRANSMEMBRANE IONIC CURRENTS: Cl⁻ CURRENT

Extracellular ATP (5-50 µM) activated an outwardly rectifying, time dependent Cl⁻ current in single guinea pig atrial myocytes; ADP, AMP and adenosine also activated this current.[75] A Cl⁻ current was activated by extracellular ATP (5×10^{-7}-10^{-4} M) also in single rat ventricular myocytes.[75] This current, blocked by the chloride channel blocker, 4-4'diisothiocyanatostilbene-2,2'-disulfonic acid (DIDS), was not activated by either AMP or adenosine. The differential action of adenosine on Cl⁻ currents, activated by extracellular ATP in atrial and ventricular myocytes could reflect different Cl⁻ channels in these cells or species variability.

It has been proposed that the activation of the Cl⁻ current at the resting potential could lead to the depletion of intracellular Cl⁻ which can activate the HCO_3-Cl⁻ exchanger leading to intracellular acidification.[76] Indeed, similar potency and efficacy of ATP was noted on its activation of Cl⁻ current[76] and intracellular acidification.[51]

The physiological importance of ATP's action on Cl⁻ current is not determined, however, it should be mentioned here that the opposite action, i.e., the attenuation of isoprenaline-induced Cl⁻ current in guinea pig ventricular myocytes has also been reported.[77]

ATP MODULATES INTRACELLULAR PH

Several studies have demonstrated an effect of extracellular ATP on transmembrane ionic exchange and transport and thereby on intracellular pH (pH_i). MgATP added to a suspension of rat cardiac cells induced a transient acidification followed by alkalinization.[78] The alkalinization seemed to be due to the activation of Na^+/H^+ antiport[78,79] as well as the Na^+/HCO_3^- cotransport.[80] A different study in isolated guinea pig ventricular myocytes has shown that extracellular ATP stimulates Na^+-H^+ exchange but inhibits Na^+-HCO_3^- cotransport such that the overall effect was a slowing of net acid extrusion from the cells.[81] The acidification, suggested to be caused by the MgATP-stimulated Cl⁻-HCO_3^- exchange,[78] was evoked also by Mg salts of ATPγS, α,βmATP, β,γmATP and β,γimidoATP but not by CTP, GTP, ITP, UTP, ADP and adenosine.[51] The signal transduction mediating this phenomenon seems to involve tyrosine phosphorylation of the exchanger.[82] Similarly, it was shown that the ATP-dependent increase in intracellular level of inositol trisphosphate (IP_3) results from the activation of phospholypase Cγ following its rapid and reversible phosphorylation at a tyrosine site[83] (Fig. 6.2).

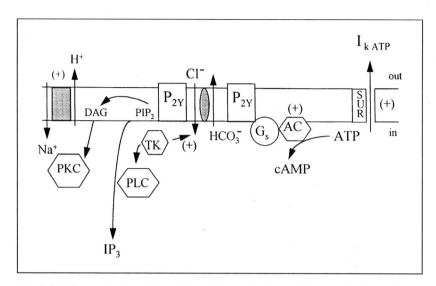

Fig. 6.2. Schematic description of signal transduction pathways which have been suggested to mediate some of the effects of extracellular ATP or cardiac myocytes. On the right, a P_{2Y} purinoceptor is linked to adenylyl cyclase (AC) via G_s protein; its activation can reduce locally the intracellular level of ATP and the activation of the ATP-sensitive K^+ channel which is blocked by sulfonyeurea (SUR) derivatives. On the left is a P_{2Y} purinoceptor whose activation stimulates tyrosine kinases which phosphorylate and activate the Cl^-/HCO_3^- exchanger and phospholipase Cγ (PLCγ). The latter leads to increased hydrolysis of phosphatidyl inositol biphosphate (PIP_2) and thereby the production of diacylglycerol (DAG) and inositol triphosphate (IP_3). The former activates protein kinase C (PKC) which could activate the Na^+-H^+ exchanger; the latter enhances the release of Ca^{2+} from internal stores (shown in Fig. 6.1).

It has been proposed that MgATP-induced intracellular acidification could lead to the following chain of events: The increase in intracellular H^+ level displaces Ca^{2+} from internal binding sites; these Ca^{2+} open a nonspecific membrane conductance which causes cell membrane depolarization and the activation of voltage dependent L-Ca^{2+} channels; this leads to Ca^{2+}-dependent Ca^{2+}-release from the SR with possible participation of increased phosphoinositide turnover.[51,59]

ARRHYTHMOGENIC EFFECTS OF ATP

In guinea pig isolated ventricular myocytes ATP alone did not exert any significant electrophysiologic effect. However, when it was applied with drugs known to increase $[Ca^{2+}]_i$, ATP facilitated the induction of afterdepolarizations and triggered activity in ~60% of the cells.[84] In the presence of isoproterenol, ATP increased the amplitude of the transient inward current (I_{ti}), delayed afterdepolarizations (DADs) and I_{Ca}; and in the presence of either BayK8644 or quinidine, ATP further prolonged the action potential duration and also increased the amplitude of early afterdepolarizations (EADs).[84] These findings extend earlier, similar observations regarding the interaction between the effects of catecholamines and ATP on I_{Ca}[49,64] and support the hypothesis that the release of ATP into the extracellular space under pathophysiologic conditions could be arrhythmogenic.[59]

In contrast to its effect on rat and guinea pig ventricular myocytes, extracellular ATP inhibited, in a concentration-dependent manner, I_{Ca} in guinea pig single sinoatrial nodal (SAN) cells.[84] The rank order of potency of ATP and related compounds in inhibiting I_{Ca} in SAN cells was: ATP = α,βmATP>>2mSATP≥ ATPγS>>UTP=AUP>AMP≥Adenosine.[85] This potency order has not been reported with regard to previously identified P_2-purinoceptor subtypes, suggesting the mediation by a novel receptor of ATP's action. In view of the critical role of I_{Ca} in the genesis of the action potential in SAN cells, it has been proposed that extracellular ATP may play an important role in the regulation of heart rate.[85]

In single cells isolated from the rabbit sino-atrial node, extracellular ATP activated a time-independent, weakly inwardly rectifying current, which was nonselective for monovalent cations.[86] This current was not activated by either ADP, AMP or adenosine suggesting that the action of ATP was mediated by a P_2-purinoceptor.

Extracellular ATP caused cardiac acceleration in 40% of the rabbit hearts studied by Takikawa et al[87] using the Langendorff perfusion method. This effect of ATP was not antagonized by either theophylline, a nonselective adenosine receptor block, or PTX pretreatment, but was almost completely and partially blocked by apamin and neomycin, a phospholipase C (PLC) inhibitor, and indomethacin, respectively.[87] Based on these data it was concluded that the positive chronotropic effect of ATP was mediated by P_2-purinoceptors coupled to prostaglandins synthesis via a PTX-insensitive pathway involving the stimulation of PLC.[87] It is difficult to extrapolate data obtained in vitro[85,87] to the human heart in vivo. However, it should be noted that numerous studies in cats, dogs and human subjects have indicated that, at least in these species, extracellular ATP exerts a negative chronotropic action on cardiac pacemakers which is mediated by the vagus nerve and/or adenosine, the product of ATP's enzymatic degradation.[88,89] The above mentioned studies strongly suggest that extracellular ATP can directly affect ionic currents in sino-atrial cells. However, to what extent this action reflects a physiological role of extracellular ATP in the regulation of sino-atrial nodal automaticity, remain to be determined.

ATP AND TRANSMEMBRANE IONIC CURRENTS: SUMMARY

The data briefly reviewed above indicate that extracellular ATP can modulate several ionic permeabilities of the cardiac myocyte membrane. Figure 6.3 summarizes the transmembrane ionic currents affected by extracellular ATP. As can be seen, ATP can induce inward depolarizing currents (i.e., I_{Na}, I_{CaT}, I_{CaL} and I_{ATP}) as well as outward repolarizing currents (i.e., I_{Kdel}, I_{KAch}, I_{KATP}, I_{Cl} and I_{ATP}). Since the relevant data were obtained under specific experimental condition in vitro, it is not known to what extent these effects play in the inotropic and electrophysiologic actions of extracellular ATP in the mammalian heart in situ. Thus, more studies are required before the role played by extracellular ATP in cardiac physiology and pathophysiology is fully determined.

EFFECTS OF ATP ON PROSTAGLANDINS SYNTHESIS AND RELEASE

Extracellular ATP stimulated the de novo synthesis and release of prostaglandins (PG) in the perfused rabbit heart.[90] In the

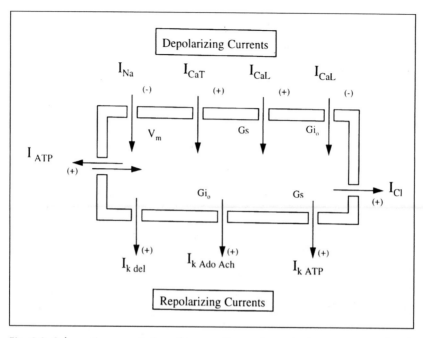

Fig. 6.3. Schematic presentation of trans-cell membrane ionic currents regulated, in part, by purine nucleosides and nucleotides. Vm = Cell membrane potential; G_s and $G_{i/o}$ are G proteins which mediate stimulation and inhibition of adenylyl cyclase respectively; I_{ATP} = Nonselective cationic current; I_{Na} = Sodium current; I_{CaT} and I_{CaL} are T and L type calcium currents, respectively; I_{Cl} = Chloride current; I_{Kdel} = Delayed outwardly rectifier potassium current and I_{KATP} = ATP-sensitive potassium current; + and - denote a stimulatory or an inhibiting effect, respectively.

latter study the site of action of ATP was not determined, however, subsequent studies have shown that extracellular ATP can induce the release of PG from cultured porcine endothelial cells[91] and bovine aortic smooth muscle cells.[92] In cultured calf endothelial cells isolated from the thoracic aorta, extracellular ATP increased the intracellular level of free Ca^{2+} ($[Ca^{2+}]_i$).[93] Similar results were obtained in cultured pig aortic endothelial cells[94] and human endothelial cells isolated from umbilical veins.[95] The effects of extracellular ATP and its analogs on $[Ca^{2+}]_i$ and the release of PGI_2 was characterized by a similar rank order of potency corresponding to mediation by P_{2Y}-purinoceptors[95] as was previously suggested.[94] Thus, it seems that the effect of extracellular ATP on the synthesis and release of PG in the heart is mediated by P_2-purinoceptors on endothelial and/or smooth muscle cells of the coronary vasculature and not the cardiac myocytes.

INTRACELLULAR ATP

Intracellular ATP is also known to affect trans-cell membrane ionic conductances.[38,96,97] However, discussion of these effects of intracellular ATP is beyond the scope of this chapter which deals solely with the effects of extracellular ATP.

INDIRECT ACTIONS OF ATP

In addition to its direct actions on cardiac myocytes, extracellular ATP can indirectly affect these cells via at least two mechanisms: One is the degradation of ATP by ectoenzymes to adenosine, and the other, the interaction of extracellular ATP with the autonomic nervous system. Regarding the former mechanism, it is expected that ATP would exert all of adenosine's actions on cardiac myocytes, provided that sufficient amounts of the nucleotide are available and the enzymatic degradation is not interfered with. Several major aspects of adenosine's actions on cardiac myocytes are discussed in other chapters in this volume and additional information can be found in several reviews published recently.[98,99]

The interactions of extracellular ATP with the autonomic nervous system has been discussed in detail in two recent reviews.[100,101] Therefore, only a brief outline of this interaction is given below.

Early anecdotal observations have suggested that the vagus nerve is involved in the negative chronotropic and dromotropic actions of ATP on sinus mode automaticity and AV nodal conduction, respectively.[88] A systematic study in a canine in model vivo, has confirmed these early observations and has indicated that the vagal component in the ATP's actions is responsible for the differential potency of equimolar doses of ATP and adenosine.[102]

In recent years several studies have shed light on the mechanism of ATP's enhancement of vagal efferent traffic to the heart. Specifically, data obtained in the dog have indicated that ATP can trigger a cardio-cardiac vagal depressor reflex by activating chemosensitive vagal afferent terminals in the heart.[103,104] This was based on the facts that relatively small amounts of ATP (0.1-1 μmol/kg) administered into the left main coronary artery caused a marked transient prolongation of sinus cycle length (SCL) which was abolished by bilateral cervical vagotomy.[104] In addition, adenosine administered in the same manner was without effect while the potency rank order of ATP and related nucleotides was: α,β methylene-ATP>2 methylthio-ATP>ATP≥β,γ methylene-ATP>ADP>>AMP=0. Furthermore, the P_{2x}-purinoceptor antagonist, pyridoxal-phosphate-6-azophenyl-2',4'-disulfonic acid, markedly attenuated the effects of ATP and the other purine nucleotides. It was concluded that ATP stimulates afferent nerve terminals in the left ventricle by activating P_{2x}-purinoceptors.[104]

The ability of ATP to trigger a cardio-cardiac vagal depressor reflex is in agreement with several previous studies which have documented the modulatory effect of extracellular ATP on the activity of intrinsic cardiac neurons.[105,106] In cultured neurons dissociated from rat parasympathetic cardiac ganglia, ATP evoked an inward current.[106] Similarly, ATP and related compounds modulated the spontaneous activity of intrinsic cardiac neurons in situ,[107] and localized application of ATP at epicardial sites elicited afferent traffic in vagal C fibers which was recorded in the nodose ganglia in the dog.[108]

It has been suggested that ATP can modulate neurotransmission release from adrenergic[109] and nonadrenergic noncholinergic[110] nerve terminals in the mammalian heart. However, the physiological importance of these phenomena remains to be determined.

The signal transduction pathways which mediate the actions of ATP on nerve endings within the heart and intrinsic cardiac neurons have not been fully determined. However, several studies have implicated P_{2x}-purinoceptors in the actions of ATP on central and peripheral neurons and nerve fibers.[111-113] P_{2x} designate a family of ligand binding channels whose activation does not involve either G proteins or intracellular second messengers.[114]

Summary

The data reviewed above indicate that extracellular ATP directly modulates trans-cell membrane ionic currents and thereby intracellular ionic content of cardiac myocrytes, and that this action is mediated by cell surface P_2-purinoceptors. In addition, ATP can trigger a cardio-cardiac, vagally mediated, depressor reflex which is manifested in bradyarrhhythmias. The exact physiological role of these actions of ATP is currently not know. However, since ATP is released from ischemic cardiac myocytes, it could be hypothesized that endogenous ATP modulates cardiac function under pathophysiological conditions and plays a mechanistic role in atropine-sensitive bradyarrhythmias associated with certain types of acute myocardial infarction.

Thus, further studies aimed at establishing the role of endogenous ATP in modulating cardiac function in general and cardiac electrophysiology in particular are warranted.

Acknowledgment

We thank Ms. Cheryl A. Council for her assistance in the preparation of this manuscript. Original studies were supported in part by grants from the NIH (#43006) and the American Heart Association, Southeastern Pennsylvania Affiliate.

References

1. Drury AB, Szent-Gyorgyi A. The physiological activity of adenine compounds with a special reference to their action upon the mammalian heart. J Physiol (Lond) 1929; 68:213-237.
2. Gordon JL. Extracellular ATP: effects, sources and fate. Biochem J 1986; 233:309-319.
3. Olsson RA, Pearson JD. Cardiovascular Purinoceptors. Physiol Rev 1990; 70:761-849.
4. Dubyak GR, El-Moatassim C. Signal transduction via P_2-purinergic receptors for extracellular ATP and other nucleotides. Am J Physiol 1993; 265:C577-C606.
5. Mills DCB, Robb IA, Roberts GCK. The release of nucleotides, 5-hydroxytryptamine and enzymes from human blood platelets during aggregation. J Physiol 1968; 195:715-729.
6. Day HJ, Holmsen H. Concepts of the blood platelet release reaction. Semin Hematol 1971; 4:3-27.

7. Holmsen H. Platelet metabolism and activation. Semin in Hematol 1985; 22:219-240.

8. Dean BM, Perrett D. Studies on adenine and adenosine metabolism by intact human erythrocytes using high performance liquid chromatography. Biochem Biophys Acta 1976; 437:1-15.

9. Bergfeld GR, Forrester T. Release of ATP from human erythrocytes in response to a brief period of hypoxia and hypercapnia. Cardiovasc Res 1992; 26:40-47.

10. Ellsworth ML, Forrester T, Ellis CG et al. The erythrocyte as a regulator of vascular tone. Am J Physiol 1995; 269:H2155-H2161.

11. Pearson JD, Gordon JL. Vascular endothelial and smooth muscle cells in culture selectively release adenine nucleotides. Nature 1979; 281:384-386.

12. Bodin P, Bailey D, Burnstock G. Increased flow-induced ATP release from isolated vascular endothelial cells but not smooth muscle cells. Br J Pharmacol 1991; 103:1203-1205.

13. Ralevic V, Milner P, Kirkpatrick KA et al. Flow-induced release of adenosine 5'-triphosphate from endothelial cells of the rat mesenteric arterial bed. Experientia 1992; 48:31-34.

14. Yang S, Cheek DJ, Westfall DP et al. Purinergic axis in cardiac blood vessels. Agonist-mediated release of ATP from cardiac endothelial cells. Circ Res 1994; 74:401-407.

15. Bodin P, Burnstock G. ATP-stimulated release of ATP by human endothelial cells. J Cardiovasc Pharmacol 1996; 27:872-875.

16. Katsuragi T, Tokunaga T, Ogawa S et al. Existence of ATP-evoked ATP release system in smooth muscles. J Pharmacol Exp Ther 1991; 259:513-518.

17. Katsuragi T, Soejima O, Tokunuga T et al. Evidence of post-junctional release of ATP evoked by stimulation of muscarinic receptors in ileal longitudinal muscles of guinea pig. J Pharmacol Exp Ther 1992; 260:1309-1313.

18. Lundberg JM. Pharmacology of co-transmission in the autonomic nervous system: integrative aspects of amines, neuropeptides, adenosine triphosphate, amino acids and nitric oxide. Pharmacol Rev 1996; 48:113-178.

19. Forrester T. An estimate of adenosine triphosphate release into the venous effluent from exercising human forearm muscle. J Physiol 1972; 244:611-628.

20. Fredholm BB, Hedqvist P, Lindstrom K et al. Release of nucleosides and nucleotides from the rabbit heart by sympathetic nerve stimulation. Acta Physiol Scand 1982; 116:285-295.

21. Darius H, Stahl GL, Lefer AM. Pharmacologic modulation of ATP release from isolated rat hearts in response to vasoconstrictor stimuli using a continuous flow technique. J Pharmacol Exp Ther 1987; 240:542-547.

22. Katsuragi T, Tokunaga T, Ohba M et al. Implication of ATP released from atrial, but not papillary, muscle segments of guinea pig by iso-proterenol and forskolin. Life Sci 1993; 53:961-967.

23. Tokunaga T, Katsuragi T, Sato C et al. ATP release evoked by iso-prenaline from adrenergic nerves of guinea pig atrium. Neurosci Let 1995; 186:95-98.

24. Paddle BM, Burnstock G. Release of ATP from perfused heart during coronary vasodilatation. Blood Vessels 1974; 11:110-119.

25. Forrester T, Williams CA. Release of adenosine triphosphate from isolated adult heart cells in response to hypoxia. J Physiol (Lond) 1977; 268:371-390.

26. Williams CA, Forrester T. Possible source of adenosine triphosphate released from rat myocytes in response to hypoxia and acidosis. Cardiovasc Res 1983; 17:301-312.

27. Al-Awqati Q. Regulation of ion channels by ABC transporters that secrete ATP. Science 1995; 269:805-806.

28. Reisin IL, Prat AG, Abraham EH et al. The cystic fibrosis transmembrane conductance regulator is a dual ATP and chloride channel. J Biol Chem 1994; 269:20584-20591.

29. Prat AG, Reisin IL, Ausiello DA et al. Cellular ATP release by the cystic fibrosis transmembrane conductance regulator. Am J Physiol (Lond) 1996; 270:C538-C545.

30. Reddy MM, Quinton PM, Haws C et al. Failure of the cystic fibrosis transmembrane conductance regulator to conduct ATP. Science 1996; 271:1876-1879.

31. Friel DD, Bean BP. Two ATP-activated conductances in bullfrog atrial cells. J Gen Physiol 1988; 91:1-27.

32. Friel DD, Bean BP. Dual control by ATP and acetylcholine of inwardly rectifying K^+ channels in bovine atrial cells. Pflügers Arch 1990; 415:651-657.

33. Fu C, Pleumsamran A, Oh U et al. Different properties of the atrial G protein-gated K^+ channels activated by extracellular ATP and adenosine. Am J Physiol 1995; 269:H1349-H1358.

34. Matsuura H, Sakaguchi M, Tsuruhara Y et al. Activation of the muscarinic K^+ channel by P_2-purinoceptors via pertussis toxin-sensitive G proteins in guinea-pig atrial cells. J Physiol (Lond) 1996; 490:659-671.

35. Hara Y, Nakaya H. Dual effects of extracellular ATP on the muscarinic acetylcholine receptor-operated K^+ current in guinea pig atrial cells. Eur J Pharmacol 1997; 324:295-303.

35a. Matsuura H, Ehara T. Modulation of the muscarinic K^+ channel by P_2-purinoceptors in guinea pig atrial myocytes. J Physiol (Lord) 1996; 497:379-393.

36. Froldi G, Pandofolo L, Cinellato A et al. Dual effect of ATP and UTP on rat atria: which type of receptors are involved? Naunyn-Schmiedeberg's Arch Pharmacol 1994; 394:381-386.

37. Matsuura H, Sakaguchi M, Tsuruhara Y et al. Enhancement of delayed rectifier K^+ current by P_2-purinoceptor stimulation in guinea pig atrial cells. J Physiol (Lond) 1996; 490:647-658.

37a. Matsuura H, Ehara T. Selective enhancement of the slow component of delayed rectifier K^+ current in guinea pig atrial cells by external ATP. J Physiol (Lond) 1997; 503:45-54.

38. Noma A. ATP-regulated K^+ channels in cardiac muscle. Nature 1983; 305:147-148.

39. Grover GJ, Dzwonczyk S, Parham CS et al. The protective effects of cromakalim and pinacidil on reperfusion function and infarct size in isolated perfused rat hearts and anesthetized dogs. Cardiovasc Drugs Ther 1990; 4:465-474.

40. Gross GJ, Auchampach JA. Blockade of ATP-sensitive potassium channels prevents myocardial preconditioning in dogs. Circ Res 1992; 70:223-233.

41. Kirsch GE, Codina J, Birnbaumer L et al. Coupling of ATP-sensitive K^+ channels to A_1 receptors by G proteins in rat ventricular myocytes. Am J Physiol 1990; 259:H820-H826.

42. Ito H, Tung RT, Sugimoto T et al. On the mechanism of G Protein $\beta\gamma$ subunit activation of the muscarinic K^+ channel in guinea pig atrial cell membrane. J Gen Physiol 1992; 99:961-983.

43. Xu J, Wang L, Hurt CM et al. Endogenous adenosine does not activate ATP- sensitive K^+ channels in the hypoxic guinea-pig ventricle in vivo. Circulation 1994 89:1209-1216.

44. Inagaki N, Goroi T, Clement JP et al. A family of sulfonylurea receptors determines the pharmacological properties of ATP- sensitive K^+ channels. Neuron 1996; 16:1011-1017.

45. Babenko AP, Vassort G. Purinergic facilitation of ATP-sensitive potassium current in rat ventricular myocytes. Br J Pharmacol 1997; 120:631-638.

46. Babenko AP, Vassort G. Enhancement of the ATP-sensitive K^+ current by extracellular ATP in rat ventricular myocytes. Involvement of adenylyl cyclase-induced subsarcolemmal ATP depletion. Circ Res 1997; 80:589-600.

47. Goto M, Yatani A, Tsuda Y. An analysis of the action of ATP and related compounds on membrane current and tension components in bullfrog atrial muscle. Jpn J Physiol 1977; 27:81-94.

48. Yatani A, Goto M, Tsuda Y. Nature of catecholamine-like actions of ATP and other energy rich nucleotides on the bullfrog atrial muscle. Jpn J Physiol 1978; 28:47-61.

49. De Young MB, Scarpa A. Extracellular ATP induces Ca^{2+} transients in cardiac myocytes which are potentiated by norepinephrine. FEBS Lett 1987; 223:53-58.

50. Sharma VK, Sheu S-S. Micromolar extracellular ATP increases intracellular calcium concentration in isolated rat ventricular myocytes (Abs.). Biophys J 1986; 49:351a.

51. Pucéat M, Clément O, Scamps F et al. Extracellular ATP-induced acidification leads to cytosolic calcium transient rise in single rat cardiac myocytes. Biochem J 1991; 274:55-62.

52. Danziger RS, Raffaeli S, Moreno-Sanchez R et al. Extracellular ATP has a potent effect to enhance cystolic calcium and contractility in single ventricular myocytes. Cell Calcium 1988; 9:193-199.

53. De Young MB, Scarpa A. ATP receptor-induced Ca^{2+} transients in cardiac myocytes: sources of mobilized Ca^{2+}. Am J Physiol 1989; 257:C750-C758.

54. Björnsson OG, Monck JR, Williamson JR. Identification of P_{2Y} purinoceptors associated with voltage-activated cation channels in cardiac ventricular myocytes of the rat. Eur J Biochem 1989; 186:395-404.

55. Christie A, Sharma VK, Sheu S-S. Mechanism of extracellular ATP-induced increase of cytosolic Ca^{2+} concentration in isolated rat ventricular myocytes. J Physiol 1992; 445:369-388.

56. Alvarez JL, Mongo K, Scamps F et al. Effects of purinergic stimulation on the Ca current in single frog cardiac cells. Pflügers Arch 1990; 416:189-195.

57. Scamps F, Legssyer A, Mayoux E et al. The mechanism of positive inotropy induced by adenosine triphosphate in rat heart. Circ Res 1990; 67:1007-1016.

58. Scamps F, Legssyer A, Mayoux E et al. A cholera toxin sensitive G-protein mediates the P_2-purinergic stimulation of calcium current in rat ventricular cells (Abs.). J Physiol (Lond), 1991.

59. Scamps F, Vassort G. Mechanism of extracellular ATP-induced depolarization in rat isolated ventricular cardiomyocytes. Pflügers Arch 1990; 417:309-316.

60. Scamps F, Vassort G. Pharmacological profile of the ATP-mediated increase in L- type calcium current amplitude and activation of a non-specific cationic current in rat ventricular cells. Br J Pharmacol 1994; 113:982-986.

61. Scamps F, Rybin V, Puceat M et al. A G_s protein couples P_2-purinergic stimulation to cardiac Ca channels without cyclic AMP production. J Gen Physiol 1992; 100:675-701.

62. Alvarez JL, Vassort G. Properties of the low threshold Ca current in single frog atrial cardiomyocytes. A comparison with the high threshold Ca current. J Gen Physiol 1992; 100:519-545.

63. De Young MB, Scarpa A. Extracellular ATP activates coordinated Na^+, P_i, and Ca^{2+} transport in cardiac myocytes. Am J Physiol 1991; 260:C1182-C1190.

64. Zheng J-S, Christie A, De Young MB et al. Synergism between cAMP and ATP in signal transduction in cardiac myocytes. Am J Physiol 1992a; 262:C128-C135.

65. Zheng J-S, Christie A, De Young MB et al. Ca^{2+} mobilization by extracellular ATP in rat cardiac myocytes: regulation by protein kinase C and A. Am J Physiol 1992; 262:C933-C940.

66. Zheng J-S, Christie A, Levy MN et al. Modulation by extracellular ATP of two distinct currents in rat myocytes. Am J Physiol 1993; 264:C1411-C1417.

67. Hirano Y, Abe S, Sawanobori T et al. External ATP-induced changes in $[Ca^{2+}]_i$ and membrane currents in mammalian atrial myocytes. Am J Physiol 1991; 260:C673-C680.

68. Qu Y, Himmel HM, Campbell DL et al. Effects of extracellular ATP on I_{Ca}, $[Ca^{2+}]_i$, and contraction in isolated ferret ventricular myocytes. Am J Physiol 1993; 264:C702-C708.

69. Qu Y, Himmel HM, Campbell DL et al. Modulation of L-type Ca^{2+} current by extracellular ATP in ferret isolated right ventricular myocytes. J Physiol (Lond) 1993; 471:295-317.

70. Qu Y, Himmel HM, Campbell DL et al. Modulation of basal L-type Ca^{2+} current by adenosine in ferret isolated right ventricular myocytes. J Physiol (Lond) 1993; 471:269-293.

71. Ito H, Hosoya Y, Inanobe A et al. Acetylcholine and adenosine activate the G protein-gated muscarinic K^+ channel in ferret ventricular myocytes. Naunyn-Schmiedeberg's Arch Pharmacol 1995; 351:610-617.

72. Legssyer A, Poggioli J, Renard D et al. ATP and other adenine compounds increase mechanical activity and inositol triphosphate production in rat heart. J Physiol 1988; 401:185-199.

73. Yamada M, Hamamori Y, Akita H et al. P_2-purinoceptor activation stimulates phosphoinositide hydrolysis and inhibits accumulation of cAMP in cultured ventricular myocytes. Circ Res 1992; 70:477-485.

74. Scamps F, Vassort G. Effect of extracellular ATP on the Na^+ current in rat ventricular myocytes. Circ Res 1994; 74:710-717.

75. Matsuura H, Ehara T. Activation of chloride current by purinergic stimulation in guinea pig heart cells. Circ Res 1992; 70:851-855.

76. Kaneda M, Fukui K, Doi K. Activation of chloride current by P_2-purinoceptors in rat ventricular myocytes. Br J Pharmacol 1994; 111:1355-1360.

77. Rankin AC, Sitsapesan, Kane KA. Antagonism by adenosine and ATP of an i isoprenaline-induced background current in guinea-pig ventricular myocytes. J Mol Cell Cardiol 1990; 22:1371-1378.

78. Pucéat M, Clément O, Vassort G. Extracellular MgATP activates the Cl^-/HCO_3^- exchanger in single rat cardiac cells. J Physiol (Lond) 1991; 444:241-256.

79. Wallert MA, Fröhlich O. Na-H exchange in isolated ventricular myocytes. Biophysical J 1989; 55:287a.

80. Terzic A, Pucéat M, Clément-Chomienne O et al. Phenylephrine and

ATP enhance an amiloride insensitive bicarbonate-dependent alkalinizing mechanism in rat single cardiomyocytes. Naunyn-Schmiedeberg's Arch Pharmacol 1992; 346:597-600.

81. Lagadic-Gossmann D, Vaughan-Jones RD, Buckler KJ. Adrenaline and extracellular ATP switch between two mode of acid extrusion in the guinea-pig ventricular myocyte. J Physiol (Lond) 1992; 458:385-407.

82. Pucéat M, Cassoly R, Vassort G. Purinergic stimulation induces a tyrosine phosphorylation of a band 3-like protein in rat cardiac cells. J Physiol (Lond) 1993; 450:226P.

83. Pucéat M, Vassort G. Purinergic stimulation of rat cardiomyocytes induces tyrosine phosphorylation and membrane association of phospholipase C: a major mechanism for InsP₃ generation. Biochem J 1996; 318:723-728.

84. Song Y, Belardinelli L. ATP promotes development of afterdepolarizations and triggered activity in cardiac myocytes. Am J Physiol 1994; 267:H2005-H2011.

85. Qi A-D, Kwan YW. Modulation by extracellular ATP of L-type calcium channels in guinea-pig single sinoatrial nodal cell. Br J Pharmacol 1996; 119:1454-1462.

86. Shoda M, Hagiwara N, Kasanuki H et al. ATP-activated cationic current in rabbit sino-atrial node cells. J Mol Cell Cardiol 1997; 29:689-695.

87. Takikawa R, Kurachi Y, Mashima S et al. Adenosine-5'-triphosphate-induced sinus tachycardia mediated by prostaglandin synthesis via phospholipase C in the rabbit heart. Pflügers Arch 1990; 417:13-20.

88. Belhassen B, Pelleg A. Electrophysiologic effects of adenosine triphosphate and adenosine in the mammalian heart: Clinical and experimental aspects. J Am Coll Cardiol 1984; 4:414-24.

89. Pelleg A. Cardiac electrophysiology and pharmacology of adenosine and ATP: Modulation by the autonomic nervous system. Clin Pharmacol 1987; 27:366-72.

90. Needlman P, Minkes MS, Douglas JR. Stimulation of prostaglandin biosynthesis by adenine nucleotides. Profile of prostaglandin release by perfused organs. Circ Res 1974; 34:455-460.

91. Pearson JD, Slakey LL, Gordon JL. Stimulation of prostaglandin production through purinoceptors on cultured porcine endothelial cells. Biochem J 1983; 214:273-276.

92. Demolle D, Lagneau C, Boeynaems J-M. Stimulation of prostacyclin release from aortic smooth muscle cells by purine and pyrimidine nucleotides. Eur J Pharmacol 1988; 155:339-343.

93. Lückhoff A, Busse R. Increased free calcium in endothelial cells under stimulation with adenine nucleotides. J Cell Physiol 1986; 126:414-420.

94. Needham L, Cusack NJ, Pearson JD et al. Characteristics of the P₂ purinoceptor that mediates prostacyclin production by pig aortic endothelial cells. Eur J Pharmacol 1987; 134:199-209.

95. Carter TD, Hallman TJ, Cusack NJ et al. Regulation of P_{2y}-purinoceptor-mediated prostacyclin release from human endothelial cells by cytoplasmic calcium concentration. Br J Pharmacol 1988; 95:1181-1190.

96. Taniguchi J, Noma A, Irisawa H. Modification of the cardiac action potential by intracellular injection of adenosine triphosphate and related substances in guinea pig single ventricular cells. Cir Res 1983; 53:131-139.

97. Sugiura H, Toyama J, Tsuboi N et al. ATP directly affects junctional conductance between paired ventricular myocytes isolated from guinea pig heart. Circ Res 1990; 66:1095-1102.

98. Belardinelli L, Pelleg A, editors. Physiology and Pharmacology of Adenosine and Adenine Nucleotides: From Molecular Biology to Patient Care. Norwell, MA, U.S.A.:Kluwer Academic Publishers, 1995.

99. Belardinelli L, Shryock JC, Song Y et al. Ionic basis of the electrophysiological actions of adenosine on cardiomyocytes. FASEB J 1995; 9:359-365.

100. Pelleg A, Katchanov G, Xu J. Purinergic modulation of neural control of cardiac function. J Auton Pharmacol 1996; 16:401-405.

101. Pelleg A, Katchanov G, Xu J. Autonomic neural control of cardiac function: Modulation by adenosine and ATP. Am J Cardiol 1997; 79(12A):11-14.

102. Pelleg A, Belhassen B, Ilia R et al. Comparative electrophysiologic effects of ATP and adenosine in the canine heart: Influence of atropine, propranolol, vagotomy, dipyridamole and aminophylline. Am J Cardiol 1995; 55:571-576.

103. Katchanov G, Xu J, Hurt CM et al. Electrophysiological-anatomic correlates of adenosine 5'-triphosphate-triggered vagal reflex in the dog. III: Role of cardiac afferents. Am J Physiol 1996; 270: H1785-H1790.

104. Katchanov G, Xu J, Clay A et al. Electrophysiological-anatomic correlates of ATP-triggered vagal reflex in the dog. IV. Stimulation of left ventricular afferent nerve terminals. Am J Physiol 1997; 272:H1899-H1903.

105. Allen TGJ, Burnstock G. The actions of adenosine 5'-triphosphate on guinea-pig intracardiac neurons in culture. Br J Pharmacol 1990; 100:269-276.

106. Fieber LA, Adams DJ. Adenosine triphosphate-evoked currents in cultured neurons dissociated from rat parasympathetic cardia ganglia. J Physiol (Lond) 1991; 434:239- 256.

107. Huang MH, Sylvén C, Pelleg A et al. Modulation of in situ canine intrinsic cardiac neuronal activity by locally applied adenosine, ATP or their analogs. Am J Physiol 1993; 265:R914-R922.

108. Armour JA, Huang MA, Pelleg A et al. Responiveness of in situ canine nodose ganglion afferent neurons to epicardial mechanical or chemical stimuli. Cardiovasc Res 1994; 28:1218-1225.

109. Von Kugelgen I, Stoffel D, Starke K. P_{2x}-purinoceptor-mediated inhibition of noradrenaline release in rat atria. Br J Pharmacol 1995; 115:247-254.

110. Rubino A, Amerini S, Ledda F et al. ATP modulates the efferent function of capsaicin-sensitive neurons in guinea-pig isolated atria. Br J Pharmacol 1992; 105:516-520.

111. Edwards FA, Gibb AJ, Colquhon D. ATP receptormediated synaptic currents in the central nervous system. Nature 1992; 359:144-147.

112. Tresize DJ, Bell NJ, Kennedy I et al. Effects of divalent cations on the potency of ATP and related agonists in the rat isolated vagus nerve: implications for P_{2x}- purinoceptor classification. Br J Pharmacol 1994; 113:463-470.

113. Khakh B, Humphrey PPA, Surprenant A. Electrophysiological properties of P_{2x}- purinoceptors in rat superior cervical nodose and guinea-pig coeliac neurons. J Physiol (Lond) 1995; 484:385-395.

114. Surprenant A, Buell G, North RA. P_{2x}-receptors bring new structure to ligang-gated ion channels. Trends Pharmacol Sci 1995; 18:224-229.

115. Burnstock G, Meghji P. Distribution of P_1- and P_2-purinoceptors in the guinea-pig and frog heart. Br J Pharmac 1981; 73:879-885.

116. Moody CJ, Meghji P, Burnstock G. Stimulation of P_1-purinoceptors by ATP depends partly on its conversion to AMP and adenosine and partly on direct action. Eur J Pharmacol 1984; 97:47-54.

117. Mantelli L, Amerini S, Filippi S et al. Blockade of adenosine receptors unmasks a stimulatory effect of ATP on cardiac contractility. Br J Pharmacol 1993; 109:1268-1271.

118. Meinertz T, Nawrath H, Scholz H. Influence of cyclization and acyl substitution on the inotropic effects of adenine nucleotides. Naunyn-Schmiedeberg's Arch Pharmacol 1973; 278:165-178.

119. Hollander PB, Webb L. Effects of adenine nucleotides on the contractility and membrane potentials of rat atrium. Circ Res 1957; 5:349-353.

120. Burnstock G, Meghji P. The effect of adenyl compounds on the rat heart. Br J Pharmac 1983; 79:211-218.

121. Hohl CM, Hearse DJ. Vascular and contractile responses to extracellular ATP: Studies in the isolated rat heart. Can J Cardiol 1985; 1:207-216.

122. Fleetwood G, Gordon JL. Purinoceptors in the rat heart. Br J Pharmac 1987; 90:219-227.

123. Acierno L, Burno F, Burnstein F et al. Actions of adenosine triphosphate on the isolated cat atrium and their antagonism by acetylcholine. J Pharmacol Exp Ther 1952; 104:264-268.

Index